An Overview of Random Variables for Electrical Engineers

by Joseph M. Pereira

ISBN 1-44-995281-X

978-1-44-995281-5

Table Of Contents

Preface

This book is not intended to thoroughly teach the subject but to provide a high level overview of some of the major principles of probability theory and random variables. The material is presented in a very simplistic manner in order to help make the concepts easier to grasp and allow faster coverage of the topics for undergraduate students. The Examples and exercises have a slant toward electrical engineering situations. Therefore, the reader of this book should be familiar with digital logic, circuit analysis, basic microelectronics and calculus.

The study of probability and random variables forms the foundation for launching into the study of random processes, the study of random signals applied to linear systems, digital and analog communications, and random signals and noise. It is hoped that this book provides a lucid and sufficient overview of probability and random variables and the tools to enable the reader to move into these more advanced areas.

When studying a subject, it is often frustrating to find that there are exercises to which there are no answers. This inhibits the reader from verifying if the material has been grasped. This book addresses the issue by providing solutions to the problems so that the reader can also check where any mistakes in the problem solving process might have occurred. Additionally, two chapters (midterm and final) provide quiz questions to test the reader's cumulative knowledge of the material. Because this book provides such resources it is strongly advised to work through the exercises before looking at the solution. This will alert the reader to where there are possible misunderstandings of the material.

Finally, I would like to thank all of the people who have made this book possible. First and foremost I want to thank my parents, Jose and Lorraine Pereira for encouraging me to write this book. I also would like to thank my sister, Lorraine Larocca for proof reading the manuscript and providing constructive criticism.

Sincerely,

Joseph M. Pereira

Chapter 1   Set Theory Basics

We begin our study by looking at set theory. As we shall see in following chapters, set theory will aid us in determining probabilities.

1.1   Introduction

A set is a grouping of objects. Each individual object within a set is called an element. The elements of a set are usually enclosed within braces, {}, and separated by commas.

Example 1.1:

The following are examples of sets:

- {0, 2, 4, 6, 8} This is the set of even integers between 0 and 9 inclusive.
- {00, 01, 10, 11} This set is made up of all the possible combinations of two binary digits.
- {} This set contains nothing and is known as the empty set. The empty set is a valid set and is also represented by $\varnothing$. ∎

Example 1.2:

We can name sets as well. As an example we can call the three sets above A, B, and $\varnothing$ as follows:

- A = {0, 2, 4, 6, 8}
- B = {00, 01, 10, 11}
- $\varnothing$ = {}   ∎

Some sets may contain so many elements that we could not possibly list them all. In such cases it is convenient to list the elements in the set in terms of their properties as follows: {ζ | ζ satisfies property P}. The "|" symbol is read as "such that".

Example 1.3:

If we want to define a set, which consists of the positive integers from 3 to 340 inclusive, we would write {ζ | $3 \le \zeta \le 340$, ζ an integer}. ∎

1.2   Basic Definitions

1.2.1 Countable Sets

A set is said to be countable if its elements can be enumerated, even if the number of elements is infinite.

Example 1.4:

An example of a countable set would be $\{\zeta \mid 3 \leq \zeta \leq 340, \zeta$ an integer$\}$. An example of a countably infinite set would be $\{\zeta \mid 0 \leq \zeta \leq \infty, \zeta$ an integer$\}$. Even though the number of elements in the set is infinite we can still list them out; therefore, the set is still considered countable. ∎

1.2.2 Uncountable Sets

An uncountable set is a set in which the elements cannot be enumerated.

Example 1.5:

An example of an uncountable set would be $\{\zeta \mid 3 \leq \zeta \leq 340\}$. This set consists of the continuum of numbers from 3 to 340 and therefore, the elements cannot be listed out. It is for this reason that the set is considered to be uncountable. For instance, the element after 3 would be 3.000000.... with an infinite number of digits after the decimal. i.e., it would take an infinite number of digits to represent each element, which is impossible to list out. ∎

1.2.3 Null or Empty Set

The null or empty set, is represented by braces with nothing in between, e.g. {} or with the symbol $\varnothing$. The null set has no elements. Note that {0} is not the null set because 0 is an element.

1.2.4 Universal Set (Sample Space)

The universal set (also called the sample space) is a set that contains all of the possible elements for the situation being considered. Ω is used to represent the universal set.

Example 1.6:

If we were considering the voltage output from an op-amp, the universal set would be $\Omega = \{Vout_{min} \leq Vo \leq Vout_{max}\}$. If we were considering the output of a 3-bit digital circuit the universal set would be $\Omega = \{000, 001, 010, 011, 100, 101, 110, 111\}$. ∎

1.2.5 Elements and Subsets

As previously stated, the individual objects contained within a set are called elements. The symbols $\in$ and $\notin$ are used to indicate if an object is or is not an element of a set.

Example 1.7:

Consider the set A={3, 7, A, G, $, &}[1]. The relational equations $7 \in A$ and $2 \notin A$ are read as "7 is an element of set A" and "2 is not an element of set A" respectively. ∎

When all of the elements of a set are contained within another set, we call the former set a subset of the later set.

Example 1.8:

Consider the set H={a, b, c, d, e, f, g}. The set D={c, e, f} is an example of a subset of the set H and is written symbolically as D⊂H. ∎

Example 1.9:

Consider the following sets Ω={a}, Ω={a,b}, and Ω={a,b,c}. Let's write out their sub-sets:

Ω={a} → {}, {a}

Ω={a,b} → {}, {a}, {b}, {a,b}

Ω={a,b,c} → {}, {a}, {b}, {c}, {a,b}, {a,c}, {b,c}, {a,b,c}

Notice from above that for a set with N elements, the total number subsets that can be made from that set is 2^N. Notice also that {} is also included as a subset. ∎

1.2.6 Disjoint Sets

Disjoint sets are sets that do not have any of their elements in common.

Example 1.10:

The following two sets are disjoint because they do not have any elements in common: {1, 3, 5, 7, 9} and {0, 2, 4, 6, 8}. However, the following two sets are not disjoint because they have at least one element, the letter T, in common: {2, y, z, &, T, $} and {3, 6, 9, T, @}. ∎

---

1. Remember, a set is a collection of objects and those objects do not have to be numbers.

1.2.7 Union of Sets

The union of two or more sets is the concatenation of their elements into one large set with any duplicate elements removed.

Example 1.11:

Consider the following two disjoint sets: $\{1, 2, 3, 4\}$ and $\{7, 8, 9\}$. Their union is represented by $\{1, 2, 3, 4\} \cup \{7, 8, 9\} = \{1, 2, 3, 4, 7, 8, 9\}$. Next consider the following two sets which are not disjoint: $\{1, 2, 3, 4, 5\}$ and $\{3, 5, 8, 11\}$. Their union is $\{1, 2, 3, 4, 5\} \cup \{3, 5, 8, 11\} = \{1, 2, 3, 4, 5, 8, 11\}$. Notice that the common elements are not included twice. ■

1.2.8 Intersection of Sets

Taking the intersection of two or more sets results in a set of the elements which are common to those sets.

Example 1.12:

Consider the sets $\{1, 2, 3, 4, 5\}$ and $\{3, 5, 8, 11\}$. The intersection of these sets is $\{1, 2, 3, 4, 5\} \cap \{3, 5, 8, 11\} = \{3, 5\}$. Note that the intersection of disjoint sets is the null set $\varnothing$.■

1.2.9 Complement of a Set

The complement of a set is the set of all elements which are not included in the set. In order to take the complement of a set we need to have the sample space Ω defined. When a set is being complemented we will use a superscript C to denote that the set is being complemented.

Example 1.13:

Suppose we have $\Omega = \{0, 1, 2, 3, 4, 5, 6, 7, 8, 9\}$ and $J = \{2, 5, 6, 7\}$. Then $J^C = \{0, 1, 3, 4, 8, 9\}$.■

1.2.10 Venn Diagrams

Venn diagrams are used to graphically represent the sample space (i.e. the universal set) and the corresponding sets and subsets. This graphic representation can be useful for seeing relationships between sets.

A square box is typically used to represent the sample space while closed shapes represent sets and subsets.

Consider the following: Ω ={0, 1, 2, 3, 4, 5, 6, 7, 8, 9}, A={4, 5, 6, 7}, and B={6, 7, 8, 9}. To construct a Venn diagram we 1st draw a box to represent the universal set. We then draw a closed shape, in this case an oval, to represent A. We then draw another closed shape to represent B. We draw A and B overlapping to represent that A and B have some elements in common. The final Venn diagram is shown in Figure 1.

Ω ={0, 1, 2, 3, 4, 5, 6, 7, 8, 9}
A={4, 5, 6, 7} B={6, 7, 8, 9}

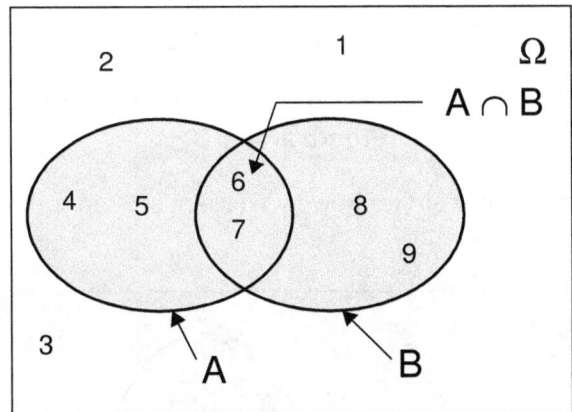

Figure 1 Venn Diagram

Note that the union of sets A and B is shown by the shaded area. The intersection of sets A and B can also be clearly seen on the diagram from where sets A and B overlap.

Below is another example Venn diagram showing set A as a subset of set B.

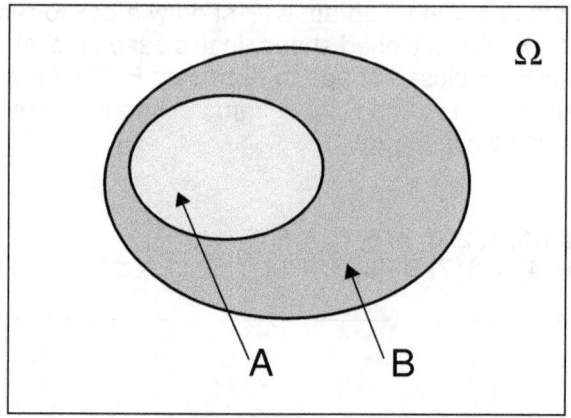

Figure 2 A ⊂ B

The Venn diagram below shows how two disjoint sets are represented.

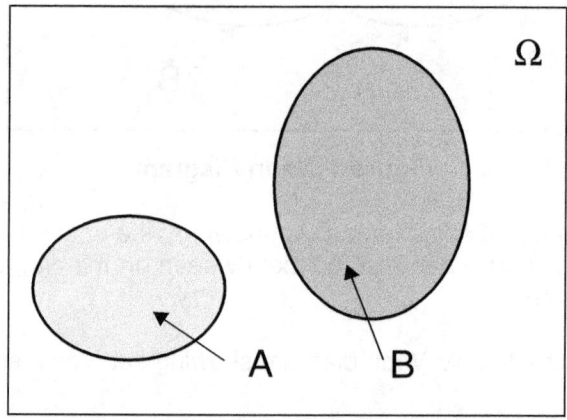

Figure 3 Sets A and B are disjoint

The Venn diagram below shows the complement of set A.

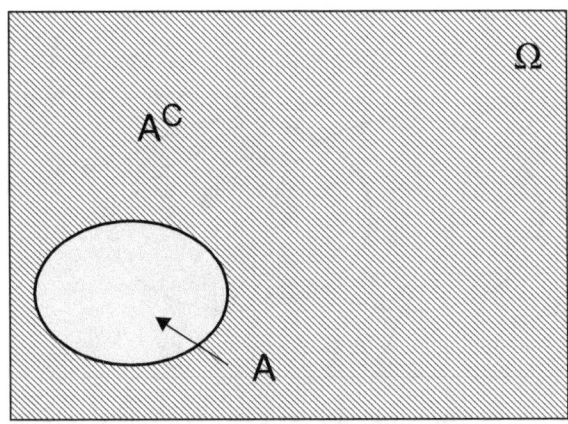

Figure 4 Complement of Set A

1.3   Set Operations

1.3.1 Adding Sets

The addition of sets is simply the union of the two sets.

If the sets are disjoint then you simply combine the elements together to make another set which is the sum.

Example 1.14:

> If we add the two disjoint sets {1, 2, 3} and {7, 9, 11, 13} we get {1, 2, 3, 7, 9, 11, 13}. ■

To add two sets which are not disjoint we combine the two sets and remove the duplicate elements.

Example 1.15:

> Consider the addition of the two sets {1, 2, 3, 4} and {1, 3, 5, 6, 7} we get {1, 1, 2, 3, 3, 4, 5, 6, 7}; then we remove the duplicate elements to get {1, 2, 3, 4, 5, 6, 7}. ■

The addition of two non-disjoint sets basically amounts to the addition of their elements minus the duplicate elements (which is the intersection of the two sets). Figure 5 below shows another way of looking at the addition of sets A and B. We make two disjoint sets, A and $A^C \cap B$ and simply add them.

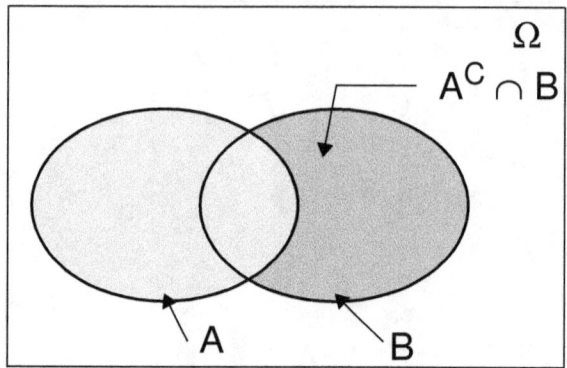

Figure 5 A ∪ B = A ∪ (A^C∩B)

1.3.2 Subtracting Sets

Subtraction of sets is accomplished by forming the intersection with the complement of the set to be subtracted. As an example, suppose we want to subtract set B from set A, i.e. A-B. We would simply create $A \cap B^C$. This can be visualized through a Venn diagram as shown below.

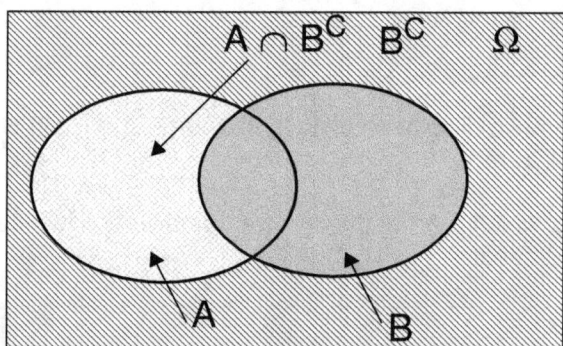

Figure 6 A - B = A ∩ B^C

1.3.3 De Morgan's Laws

De Morgan's laws are stated as follows:

[1] $(A \cap B)^C = A^C \cup B^C$

[2] $(A \cup B)^C = A^C \cap B^C$

1.3.4 Practical Set Application

Understanding how to use sets to represent real world situations will become important when we study probability. Therefore, we will look at a couple of examples to help get us warmed up.

Consider the two series switches below.

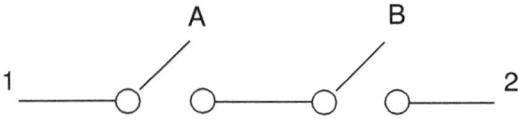

Figure 7 Series Switches

Let's say that A is the event that switch A is closed and that B is the event that switch B is closed. We can then define the event that there exists a path from point 1 to point 2 as $C_{12} = A \cap B$. This is because both switches A and B need to be closed so that there is a path between points 1 and 2. Note that the complement of C_{12} is the event that a path does not exist between points 1 and 2. Now consider the two parallel switches below.

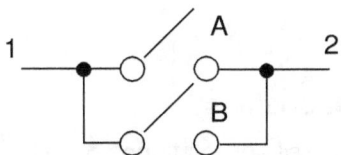

Figure 8 Parallel Switches

Let's say that A is the event that the switch A is closed and that B is the event that switch B is closed. We can then define the event that there exists a path from point 1 to point 2 as $C_{12} = A \cup B$. This is because either switch A or switch B need to be closed to make a path between points 1 and 2. Of course the complement of C_{12} is the event that a path does not exist between points 1 and 2. In subsequent chapters where we deal with the subject of independent events we will find that representing events as intersections instead of unions will make calculations easier. So by taking the complement of C_{12} we get $A^C \cap B^C$. However, to maintain the equality with the original C_{12} we need to complement the result again as follows: $C_{12} = (A^C \cap B^C)^C$.

We can combine parallel and serial switches to handle more complex situations, like the one shown in Figure 9 below.

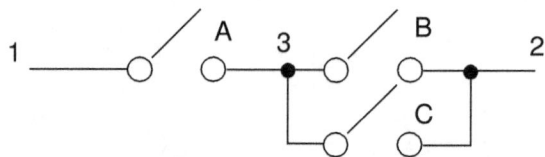

Figure 9 Series and Parallel Switches

Let's say that A, B, and C are the events that switches A, B, and C are closed respectively. The event that there is a path from point 1 to point 2 can be found by breaking up the problem into smaller problems. The event that a path exists between points 1 and 2 can be seen to be the intersection of events that a path exists between points 1 and 3 and between points 3 and 2. The event that a path exists between points 1 and 3 is simply event A. The event that a path exists between points 3 and 2 is the union of events B and C. Putting this all together gives, $D_{12} = A \cap (B \cup C)$. However we would like to use just intersections. Therefore, we convert $B \cup C$ to $(B^C \cap C^C)^C$. Then $D_{12} = A \cap (B^C \cap C^C)^C$.

1.3.5 Useful Set Relationships

Below is a list of some useful set relationships. The reader is encouraged to verify these relationships. Drawing a Venn diagram may prove useful in seeing some of these relationships.

[3] $\Omega^C = \varnothing$

[4] $\varnothing^C = \Omega$

[5] $A \cap \Omega = A$

[6] $A \cap \varnothing = \varnothing$

[7] $A \cap A^C = \varnothing$

[8] $A \cup A^C = \Omega$

[9] $A \cup \Omega = \Omega$

[10] $A \cup \varnothing = A$

[11] $A \cap B = \varnothing$, if A and B are disjoint

[12] $A \cap B = B \cap A$

[13] $A \cap (B \cap C) = A \cap B \cap C$

[14] $A \cup (B \cup C) = A \cup B \cup C$

[15] $A \cap (B \cup C) = (A \cap B) \cup (A \cap C)$

[16] $A \cup (B \cap C) = (A \cup B) \cap (A \cup C)$

[17] $A \cap A = A$

1.4   Exercises

1 Consider the following three sets: A={1, 2, 3, 6, 7}, B={2, 4, 6, 8}, and Ω={1, 2, 3, 4, 5, 6, 7, 8, 9}. Calculate the following:

a) $A \cap B$

b) $A \cup B$

c) $(A^C \cap B)^C$

d) $A \cap \Omega$

e) $A \cup \Omega$

f) $A \cap \Omega^C$

g) $A \cup \Omega^C$

2 Which of these sets is disjoint with respect to set A in exercise 1 above?

a) {5,9}

b) {}

c) $A \cap B^C$

d) A^C

e) $A \cap B$

3 Draw the Venn diagram for each of the following situations:

a) $(A^C \cap B)^C$

b) $(A \cap B)^C$

c) $A^C \cup B^C$

d) $(A \cup B)^C$

e) $A^C \cap B^C$

4 For each of the following experiments, what is the the universal set Ω (i.e. sample space)?

    a)   Three binary bits are observed periodically at the output of a digital integrated circuit.

    b)   The number of 1s observed in a randomly sampled string of seven binary bits.

    c)   The output voltage of the circuit below is observed:

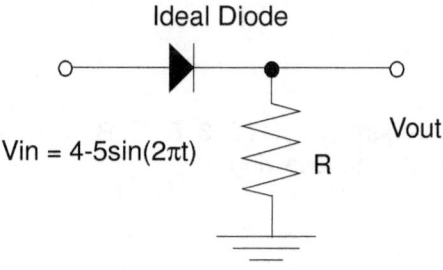

Ideal Diode

$Vin = 4-5\sin(2\pi t)$

R

Vout

Figure 10 Exercise c)

5 What are all of the possible subsets for the set {a, b, c, d}?

6 For the switch circuit below, A-D represent the event that the corresponding switch is closed. Find the event that there will be a path from point 1 to point 2. Express the final result using only intersections.

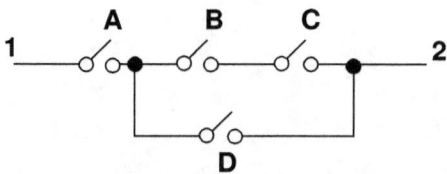

A   B   C

1   2

D

Figure 11 Switch circuit for Exercise 6

1.5   Solutions to Exercises

1

    a) $A \cap B$: The elements common to set A and set B are 2 and 6 so the answer is {2, 6}.

    b) $A \cup B$: Combining both sets gives: {1, 2, 3, 4, 6, 7, 8}

c) $(A^C \cap B)^C$: As in algebra we can do the operation within the parentheses 1st. A^C are all of the elements not contained in A. This results in A^C ={4, 5, 8, 9}. Next we find $A^C \cap B$. The elements common between these two sets are 4 and 8, therefore, $A^C \cap B = \{4, 8\}$. Finally, we take the complement of this set which is {1, 2, 3, 5, 6, 7, 9}

d) $A \cap \Omega$: The elements that are common between A and Ω are the elements in set A. Therefore, the answer is set A.

e) $A \cup \Omega$: By combining A and Ω we end up with all of the elements contained in Ω. Therefore, the answer is Ω.

f) $A \cap \Omega^C$: Ω^C is equal to the null set $\varnothing$. Therefore, the elements which are common between set A and Ω^C are nothing, i.e. the null set $\varnothing$.

g) $A \cup \Omega^C$: Ω^C is equal to the null set $\varnothing$. Combining the null set and set A results in just set A.

2

a) {5,9}: Since 5 and 9 are not in set A, this set is disjoint with respect to set A.

b) {}: This set is not disjoint from A since the null set is a subset of any set.

c) $A \cap B^C$: This set is {1, 2, 3, 6, 7} $\cap$ {1, 3, 5, 7, 9} = {1, 3, 7}. Since that set has elements which are common with set A, it is not disjoint with respect to set A.

d) A^C: This set is disjoint from set A because it has by definition no elements of set A.

e) $A \cap B$: This set is {2, 6}. Therefore, because at least one of the elements of this set are also contained in set A, this set is not disjoint with respect to set A.

3

a) $(A^C \cap B)^C$: This can be done in two steps. 1st let's draw the Venn diagram for $A^C \cap B$ which is shown below.

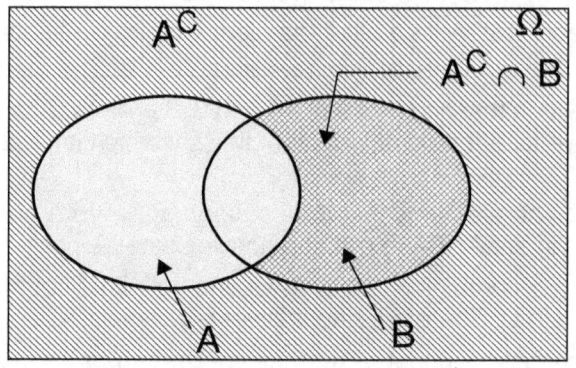

Figure 12 $A^C \cap B$

Next we take the complement of $A^C \cap B$.

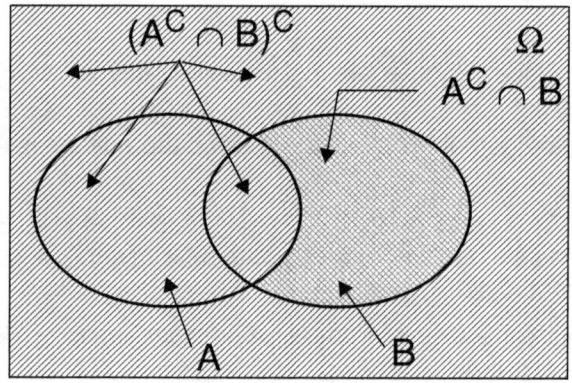

Figure 13 $(A^C \cap B)^C$

b) $(A \cap B)^C$: First we draw $A \cap B$. Then we take the complement of $A \cap B$ as shown in the figures below. $(A \cap B)^C$ is shown as the shaded area.

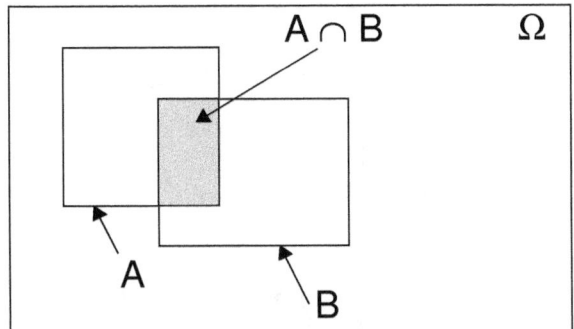

Figure 14 A ∩ B for Exercise b)

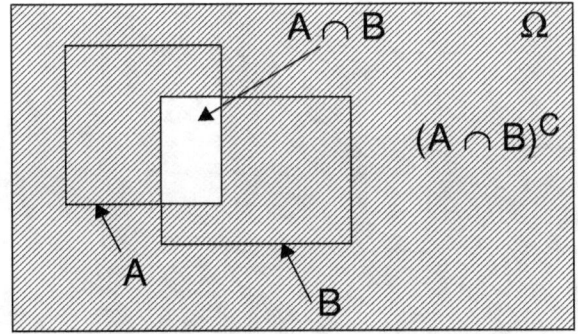

Figure 15 (A ∩ B)^C for Exercise b)

c) $A^C \cup B^C$: This can done by drawing A^C then B^C and then taking the union of the shaded areas, This is shown in the figures below.

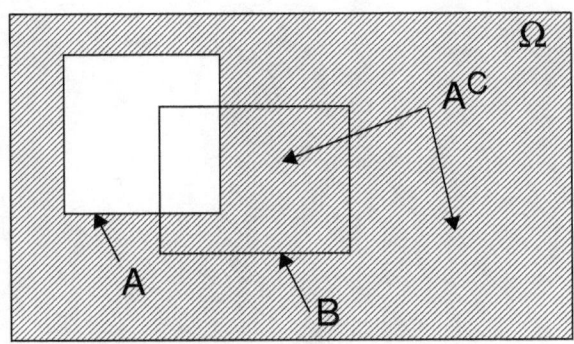

Figure 16 A^C for Exercise c)

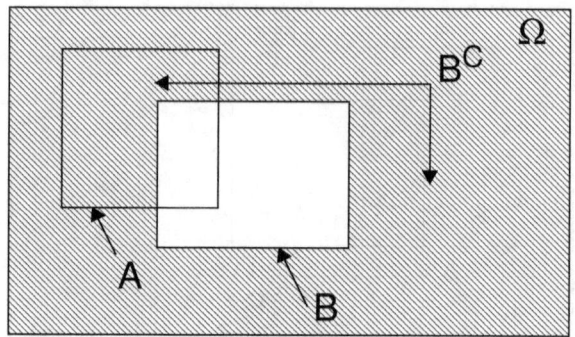

Figure 17 B^C for Exercise c)

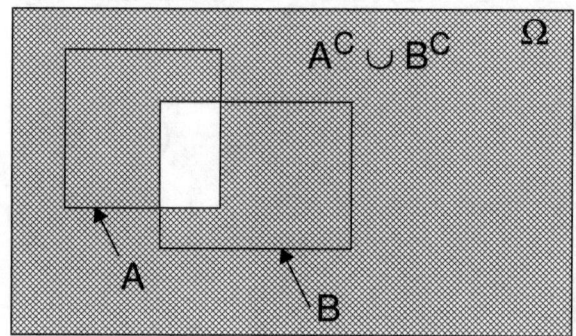

Figure 18 $A^C \cup B^C$ for Exercise c)

d) $(A \cup B)^C$: This can be easily visualized as shown in the figure below. $A \cup B$ is the shaded area. $(A \cup B)^C$ is the area not contained in $(A \cup B)$.

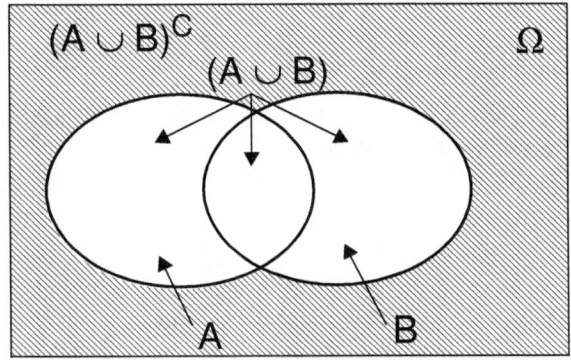

Figure 19 (A ∪ B)^C

e) $A^C \cap B^C$: This can be solved like in exercise c) except that we take the intersection of A^C and B^C. The result is given below with $A^C \cap B^C$ shown as the shaded area.

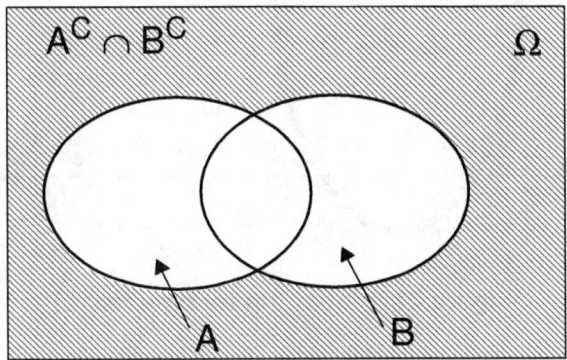

Figure 20 A^C ∩ B^C

You may have noticed that exercises b) and c) have the same graphical answer and that exercises d) and e) also have the same graphical answer. This is because these are equivalent sets, just expressed differently.

4

a) Three binary bits are observed periodically at the output of a digital integrated circuit. The sample space will consist of all the possible three-bit groups of bits that can be observed. These are: Ω = {000, 001, 010, 011, 100, 101, 110, 111}.

b) The number of 1s observed in a randomly sampled string of seven binary bits. The sample space consists of the number of 1s which can be observed in a string of 7-bits. This could range anywhere from 0 to 7. Therefore: $\Omega = \{0, 1, 2, 3, 4, 5, 6, 7\}$.

c) The output voltage of the half-wave rectifier circuit is observed. First plot or sketch the voltage Vin.

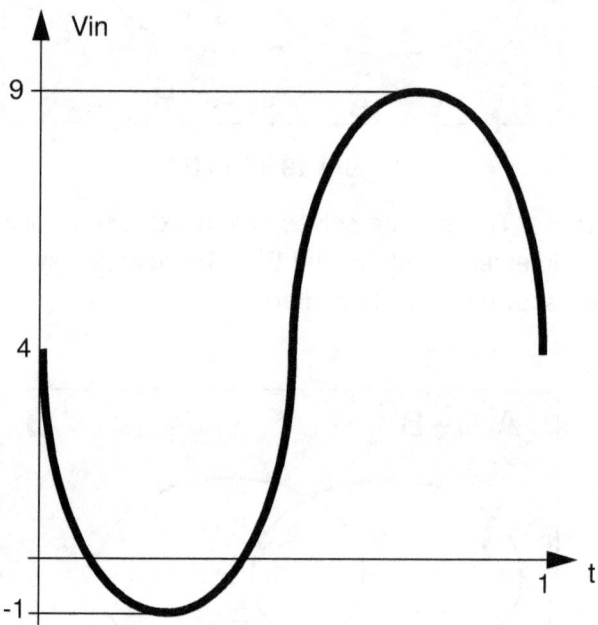

Figure 21 Plot of Vin for Exercise c)

The diode rectifies Vin so that only the portions of Vin that are greater than 0V are passed. So Vout will range from 0V to 9V. Therefore, $\Omega = \{0 \le \text{Vout} \le 9\}$.

5 All of the possible sub-sets are listed below. Note that the universal set and the null set are included as sub-sets. It is also interesting to note that the total number of subsets is 2^N, where N is the number of elements in the set.

$\{\}$, $\{a\}$, $\{b\}$, $\{c\}$, $\{d\}$, $\{a, b\}$, $\{a, c\}$, $\{a, d\}$, $\{b,c\}$, $\{b,d\}$, $\{c,d\}$, $\{a, b, c\}$, $\{a, b, d\}$, $\{b, c, d\}$, $\{a, c, d\}$, $\{a, b, c, d\}$

6 The switch circuit is shown below.

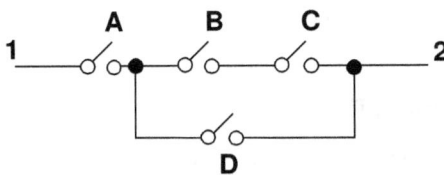

Figure 22 Switch Circuit for Exercise 6

To get a path through points 1 to 2 we need to have switch A closed and either switch D closed or switches B and C closed. This can be written as $E_{12} = A \cap [(B \cap C) \cup D]$. Now we need to turn the union into an intersection without changing the relation. By taking the complement of $(B \cap C) \cup D$ we get $(B \cap C)^C \cap D^C$. Now in order to make $(B \cap C)^C \cap D^C = (B \cap C) \cup D$ we need to add a complement to it. Therefore, we get $[(B \cap C)^C \cap D^C]^C = (B \cap C) \cup D$. Therefore $E_{12} = A \cap [(B \cap C)^C \cap D^C]^C$.

Chapter 2   Probability

In this chapter we look at some of the most basic elements of probability. A basic understanding of probability will be needed in order to understand random variables.

2.1   Relative Frequency

When we think of probability, our intuition tells us that if we repeat an experiment N times and observe an event A, n_A times, then we define the probability of event A as:

[18] $P(A) = \lim_{N \to \infty} (n_A)/N$

From the above relation we can see that $1 \geq P(A) \geq 0$ because n_A can range from 0 to N. Also, if we consider two disjoint events A and B we can see that the probability of their union is:

[19] $P(A \cup B) = \lim_{N \to \infty} (n_A + n_B)/N = \lim_{N \to \infty} (n_A)/N + \lim_{N \to \infty} (n_B)/N$

$= P(A) + P(B)$

2.2   The Axioms of Probability

Three steps that are typically followed for solving probability problems are:

- Define the sample space: i.e. this is the universal set Ω which defines all of the possible elementary outcomes of an experiment.
- Define the event space: This is all of the events of interest for the experiment.
- Assign the probability to every event in the event space.

The assignment of a probability shall follow the following axioms of probability:

[20] $1 \geq P(A) \geq 0$

[21] $P(\Omega) = 1$

[22] $P(A \cup B) = P(A) + P(B)$ if $A \cap B = \varnothing$

$$[23] \quad P\left(\bigcup_{i=1}^{\infty} A_i\right) = \sum_{i=1}^{\infty} P(A_i) \quad \text{provided that } A_i \cap A_j = \varnothing \text{ for } i \neq j.$$

[23] is just an extension of [22].

Example 2.1:

    Let's say that we conduct an experiment where we randomly sample 3-bits at the output of a binary communications channel. We define event A as observing at least two 1s. We want to find P(A). First we need to identify the sample space, which in this case is: Ω= {000, 001, 010, 011, 100, 101, 110, 111}. Next we define the event space, i.e. all of the outcomes which have at least two 1s. Therefore: A={011, 101, 110, 111}. Next, we assign a probability to seeing event A. It seems reasonable to say that since there are 4 elements in A and 8 elements in Ω then P(A)=4/8=1/2. ∎

Before moving on, we shall derive an important relationship, that is, determining $P(A^C)$ if P(A) is given. First note that $\Omega = A^C \cup A$ and that $A^C \cap A = \varnothing$. Therefore, we have $P(\Omega)=P(A^C \cup A)$, and using [21] and [22] results in 1 = $P(A^C) + P(A)$. Solving for $P(A^C)$:

[24] $P(A^C) = 1 - P(A)$.

[24] is useful when there are only two possible outcomes in an experiment, like situations where a binary digit is being observed or a switch is turned on or off. For example, in a binary system if the probability of observing a 1 is say 0.85 then the probability of observing a 0 is 0.15. Similarly, if the probability of a device being defective 0.003, then the probability of that device not being defective is 0.997.

We have just considered discrete cases where the elements in the sets are countable. When we look at continuous cases the number of elements is infinite. Let's say Ω is the real line $0 \leq x \leq 3$. If we wanted to know P(x=3.4) we would end up with 1/∞=0, due to the infinite number of elements in Ω. But we could find P(1 $\leq x \leq$ 2.5) which would be (2.5-1)/(3-0)=1/2.

2.3 Joint Probability

The joint probability of two events is defined as $P(A \cap B)$. There often exists cases where $A \cap B \neq \varnothing$. A simple way of handling this is to use [22], and the

Venn diagram below. Note that A and $A^C \cap B$ are disjoint, therefore, using [22] we can then write:

[25] $P(A \cup B) = P(A) + P(A^C \cap B)$

Now note that $A^C \cap B$ and $A \cap B$ are both disjoint so using [22] again we can write

[26] $P(B) = P(A^C \cap B) + P(A \cap B)$.

Solving [26] for $P(A^C \cap B)$ gives

[27] $P(A^C \cap B) = P(B) - P(A \cap B)$

Inserting [27] into [25] gives the result:

[28] $P(A \cup B) = P(A) + P(B) - P(A \cap B)$

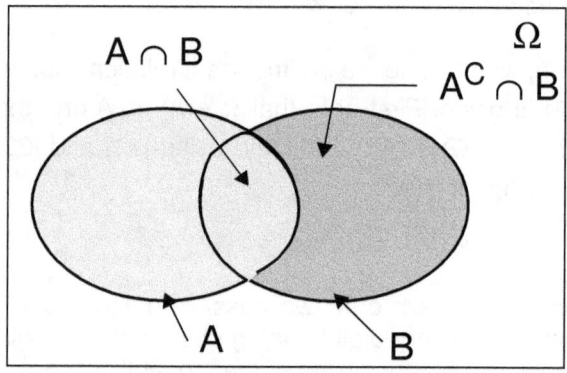

Figure 23 $A \cup B = A \cup (A^C \cap B)$

Example 2.2:

Let us return to the experiment of sampling three binary bits at the output of a binary communications channel. We define event A as observing at least two 1s. We define event B as observing a 001, 010, or 011. We want to know $P(A \cup B)$. The sample space is still $\Omega = \{000, 001, 010, 011, 100, 101, 110, 111\}$. The event space consists of the two events A=\{011, 101, 110, 111\} and B=\{001, 010, 011\}. We know from 28 that $P(A \cup B) = P(A) + P(B) - P(A \cap B)$. $P(A) = 1/2$ and $P(B) = 3/8$. $A \cap B = \{011\}$ therefore, $P(A \cap B) = 1/8$. Thus, $P(A \cup B) = 1/2 + 3/8 - 1/8 = 6/8 = 3/4$. As a check on our answer we determine $A \cup B =$

{001, 010, 011, 101, 110, 111}. Therefore, $P(A \cup B) = 6/8 = 3/4$ as expected.∎

Finally note that [28] can be used multiple times for finding the probability of the union of more than two events.

2.4 Total Probability

In the Venn diagram below, we partition the sample space in to a number of disjoint events A_1-A_{15}.

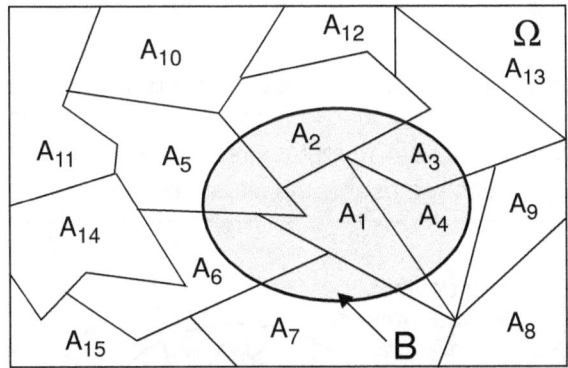

Figure 24 Venn diagram for deriving total probability

Since each $B \cap A_i$ are disjoint from each other we can determine the probability of event B by summing the probabilities of the individual $B \cap A_i$ as follows:

$$[29] \quad P(B) = \sum_{i=1}^{n} P(B \cap A_i)$$

This is the total probability for event B. n is equal to the number of events that intersect with B. In the case of Figure 24 n=7.

Example 2.3:

Manufacturer A_1 produces microchips with a probability of being defective equal to 0.001 and manufacturer A_2 produces the same microchip with a probability of being defective equal to 0.01. Now suppose that 1000 devices from manufacturer A_1 are added to a bin with 500 devices from manufacturer A_2. If a device is drawn from the bin at random, what is the probability that it is defective? First, define event B as being the event of picking a defective part. Then define A_1 as selecting a part from manufacturer A and define A_2 as selecting a part from manufacture A_2. Then using [29]:

$$P(B) = \sum_{i=1}^{n=2} P(B \cap A_i) = P(B \cap A_1) + P(B \cap A_2)$$

$$= \frac{0.001 \times 1000}{1000 + 500} + \frac{0.01 \times 500}{1000 + 500} = \frac{6}{1500} = \frac{1}{250}$$

To help clarify this example consider that out of 1000 devices from manufacturer A_1, 0.001x1000=1 device is probably in error and 0.01x500=5 devices from manufacturer A_2 are in error. Below is a Venn diagram which represents the situation.

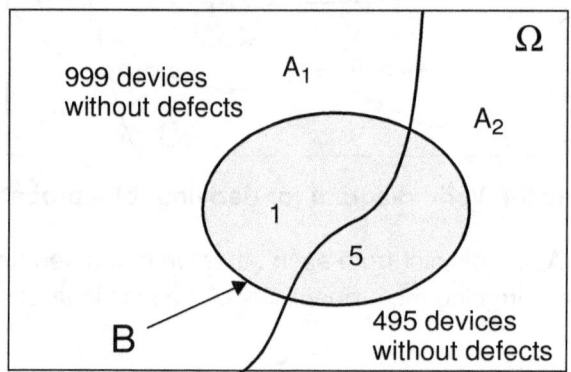

Figure 25 Venn diagram for total probability example

One can see that P(B) is:

$$\frac{\text{\# of elements in B}}{\text{\# of elements in } \Omega} = \frac{6}{1500} = \frac{1}{250}$$

This is the same result that we obtained by using [29].■

We will utilize total probability in the following sections.

2.5   Conditional Probability

There may arise cases where we wish to know the probability of an event given that another event has occurred. The concept of a conditional probability helps us find the answer.

24

Conditional probability is defined as:

$$[30] \quad P(A|B) = \frac{P(A \cap B)}{P(B)}$$

and is read as "the probability of A given that event B has occurred".

To better understand [30] take a look at the Venn diagram below.

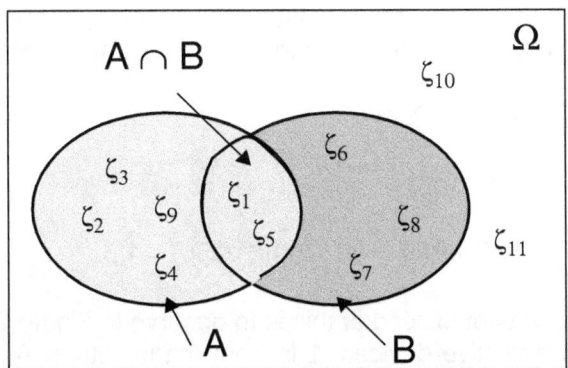

Figure 26 Venn diagram for illustrating conditional probability

We can observe that $\Omega = \{\zeta_1, \zeta_2, \zeta_3, \zeta_4, \zeta_5, \zeta_6, \zeta_7, \zeta_8, \zeta_9, \zeta_{10}, \zeta_{11}\}$, A= $\{\zeta_1, \zeta_2, \zeta_3, \zeta_4, \zeta_5, \zeta_9\}$, B= $\{\zeta_1, \zeta_5, \zeta_6, \zeta_7, \zeta_8\}$ and A $\cap$ B = $\{\zeta_1, \zeta_5\}$.

Since this is a conditional probability, and we are assuming event B has occurred, we can only consider the elements of A which are common to B, i.e., A $\cap$ B. Now observe that:

$$\frac{\# \text{ of elements in } A \cap B}{\# \text{ of elements in } \Omega} = P(A \cap B)$$

$$\frac{\# \text{ of elements in } B}{\# \text{ of elements in } \Omega} = P(B)$$

Therefore, dividing $P(A \cap B)$ by $P(B)$ we arrive at the desired result:

$$\frac{\# \text{ of elements in } A \cap B}{\# \text{ of elements in } B} = \frac{P(A \cap B)}{P(B)}$$

Example 2.4:

Let's revisit the example from the previous section. Manufacturer A_1 produces microchips with a probability of being defective equal to

25

0.001. Manufacturer A_2 produces microchips with a probability of being defective equal to 0.01. Let's say that 1000 parts from manufacturer A_1 are mixed in with 500 parts from manufacturer A_2. A device is pulled from the bin and found to be defective. What is the probability that it came from manufacturer A_1? Let's define the following:

B = {a defective part is picked}

A_1 = {manufacturer is A_1}

The joint probability $P(B \cap A_1)$ is equal to the number of elements in B $\cap A_1$ divided by the number of elements in the sample space Ω. Thus:

$P(B \cap A_1) = 1/1500$

Now we can apply [30]:

$P(A_1|B) = P(B \cap A_1)/P(B) = (1/1500)/(1/250) = 1/6$.

An alternate way of looking at this is to observe in Figure 25 that in B there are 6 defective devices. 1 is from manufacturer A_1 and 5 are from manufacturer A_2. Thus, there is a one in six chance that the defective device came from manufacturer A_1.■

2.6   Bayes' Theorem

Consider the two conditional probabilities:

[31] $P(A|B) = P(A \cap B)/P(B)$

[32] $P(B|A) = P(B \cap A)/P(A)$

We can solve [31] and 32 for the joint probabilities as follows:

[33] $P(A \cap B) = P(A|B)P(B)$

[34] $P(B \cap A) = P(B|A)P(A)$

Since $P(A \cap B) = P(B \cap A)$ we can set [33] and [34] equal to each other and thus we have:

[35] $P(A|B)P(B) = P(B|A)P(A)$

then solving [35] for the conditional probabilities we obtain the following two relationships:

$$[36] \quad P(A|B) = \frac{P(B|A)P(A)}{P(B)}$$

$$[37] \quad P(B|A) = \frac{P(A|B)P(B)}{P(A)}$$

Equations [36] and [37] allow us to determine conditional probabilities from conditional probabilities that are conditioned on the opposite event. Note that in [36] and [37] $P(B)^1$ and $P(A)$ need to be greater than 0, otherwise we would end up with a probability of infinity which of course is an invalid probability.

Equations [36] and [37] can be developed further. Note from Figure 24 that we can rewrite the conditional probability as:

[38] $P(B|A_i) = P(B \cap A_i)/P(A_i)$

Rearranging [38] into $P(B \cap A_i)=P(B|A_i)P(A_i)$, and inserting into [29] we get an additional form of the total probability theorem:

$$[39] \quad P(B) = \sum_{i=1}^{n} P(B|A_i)P(A_i)$$

Thus, inserting [38] and [39] into [36] we obtain Bayes' Theorem:

$$[40] \quad P(A_i|B) = \frac{P(B|A_i)P(A_i)}{\sum_{j=1}^{n} P(B|A_j)P(A_j)}$$

Example 2.5:

Let's look at an example. There are three bins of components A_1, A_2, and A_3. The probability that there is a defective device in the three bins is 0.01, 0.04, and 0.2 respectively. We randomly pick one of the bins, remove a component from it, and test it. We find that the compo-

---

1. P(A) and P(B) are called marginal probabilities.

nent is defective. What is the probability that it came from bin A_3? i.e. given that the device is defective what is the probability that it came from bin A_3. In [40] above B = {device is defective}. We are already given $P(B|A_1)=0.01$, $P(B|A_2)=0.04$, and $P(B|A_3)=0.2$. Because the bins were chosen at random the probability of selecting any of the bins is $P(A_1) = P(A_2) = P(A_3) = 1/3$. Therefore,

$$P(A_3|B) = \frac{P(B|A_3)P(A_3)}{P(B|A_1)P(A_1) + P(B|A_2)P(A_2) + P(B|A_3)P(A_3)}$$

$$P(A_3|B) = \frac{0.2 \times \frac{1}{3}}{0.01 \times \frac{1}{3} + 0.04 \times \frac{1}{3} + 0.2 \times \frac{1}{3}} = 0.8 \ \blacksquare$$

2.7   Independence

A very important topic in probability is independence. Two events are independent if:

[41] $P(A \cap B) = P(A)P(B)$

This can be expanded to multiple events A_1, A_2,... A_N. However, it is not enough that pair-wise independence be shown to prove independence. Every combination of joint and marginal probabilities must also be met. The following relations must be met:

[42] $P(A_i \cap A_j) = P(A_i)P(A_j)$ for all i and j

[43] $P(A_i \cap A_j \cap A_k) = P(A_i)P(A_j)P(A_k)$ for all i, j and k

...

[44] $P(A_i \cap A_j \cap ...A_N) = P(A_i)P(A_j)...P(A_N)$ for all i, j... N

If events A and B are independent then:

$P(A|B) = P(A)P(B)/P(B) = P(A)$ and $P(B|A) = P(A)P(B)/P(A) = P(B)$ which shows that if A and B are independent the occurrence of B does not affect the occurrence of A and vice versa.

Looking back at [28] if A and B are independent we get:

[45] $P(A \cup B) = P(A) + P(B) - P(A \cap B) = P(A) + P(B) - P(A)P(B)$

Example 2.6:

We are given $\Omega = \{000, 001, 010, 011, 100, 101, 110, 111\}$ and events $A=\{011, 101, 110, 111\}$, $B=\{001, 010, 011\}$, and $C=\{010, 011\}$. Are events A and B independent? We know that $P(A \cap B) = 1/8$ and $P(A)=1/2$ and $P(B)=3/8$. Therefore, since $P(A \cap B) \neq P(A)P(B)$ events A and B are not independent. Are sets A and C independent? $P(A \cap C) = 1/8$ and $P(A)=1/2$ and $P(C)=1/4$. $P(A \cap C) = P(A)P(C)$, therefore, sets A and C are independent. Now consider $P(A|C)$. $P(C) = 1/4$ and $P(A \cap C) = 1/8$. Therefore, $P(A|C) = (1/8)/(1/4) = 1/2$, which is equal to $P(A)$. This shows that indeed, the occurrence of C does not affect $P(A)$. ∎

Next, let us consider several repeated experiments such as periodically observing the output of a digital logic gate three times. We further assume that the result of the current experiment is not affected by the results of any previous experiment, thus implying independence of each experiment's results. Assuming the probability of observing a 1 is p (which means the probability of observing a 0 is 1-p), we can assign probabilities to events as follows:

$$P(0 \cap 0 \cap 0)=P(0)P(0)P(0) = p^0(1-p)^3$$

$$P(0 \cap 0 \cap 1)=P(0)P(0)P(1) = p^1(1-p)^2$$

$$P(0 \cap 1 \cap 0)=P(0)P(1)P(0) = p^1(1-p)^2$$

$$P(0 \cap 1 \cap 1)=P(0)P(1)P(1) = p^2(1-p)^1$$

$$P(1 \cap 0 \cap 0)=P(1)P(0)P(0) = p^1(1-p)^2$$

$$P(1 \cap 0 \cap 1)=P(1)P(0)P(1) = p^2(1-p)^1$$

$$P(1 \cap 1 \cap 0)=P(1)P(1)P(0) = p^2(1-p)^1$$

$$P(1 \cap 1 \cap 1)=P(1)P(1)P(1) = p^3(1-p)^0$$

Now we can find out other probabilities for this experiment such as what is the probability of seeing at least two 1s. This is found by summing the probabilities where there is at least two 1s as follows: P(2 or more ones observed) = $P(0 \cap 1 \cap 1) + P(1 \cap 0 \cap 1) + P(1 \cap 1 \cap 0) + P(1 \cap 1 \cap 1) = 3p^2(1-p)^1 + p^3(1-p)^0$.

This leads us to our next subject.

2.8   Bernoulli Trials

Bernoulli trials consist of repeating the same experiment n times and determining the probability of having a specific outcome occur a certain number of times. Only two outcomes are defined for the Bernoulli trial and the outcomes are either the event occurs or does not occur. Each outcome is assumed to be independent from the other outcomes. That is to say the current outcome does not depend on any previous outcomes and does not influence any future outcomes.

An example of a Bernoulli trial is if we were to observe one bit at the output of a binary communications channel 8 times and wanted to find the probability of seeing 3 ones.

Let's take this a step further and try to find this probability. Let's define the probability of seeing a 1 as p. Therefore, the probability of not seeing a 1 (i.e. the probability of seeing a 0) is 1-p (remember [24]). Out of 10 trials if there are three 1s then there must be seven 0s; and because each outcome is independent the probability of seeing 3 ones out of ten trials is:

[46] $(p)(p)(p)(1-p)(1-p)(1-p)(1-p)(1-p)(1-p)(1-p) = p^3(1-p)^7$

However [46] only gives one way that the three 1s and seven 0s can fall out. Some other ways are:

$(p)(1-p)(1-p)(1-p)(p)(1-p)(p)(1-p)(1-p)(1-p) = p^3(1-p)^7$

or

$(p)(p)(1-p)(p)(1-p)(1-p)(1-p)(1-p)(1-p)(1-p) = p^3(1-p)^7$

etc.

Notice that the probabilities of each instance are the same. Therefore, to find P(3 ones) we need to sum up the probabilities for all of the combinations that three 1s and seven 0s can make. The binomial coefficient[1] will allow us to do this; it is defined as:

---

1. More generally the binomial coefficient gives the number of groups each containing k items, that an be made from a group of n items. e.g., given 4 items a, b, c, and d, how many 2 item groups can be made? Grouping we get (a,b), (a,c), (a,d), (b,c), (b,d), (c,d). This is 6 items which is what the binomial coefficient predicts.

[47] $\dbinom{n}{k} = \dfrac{n!}{k!(n-k)!}$

Now we can find P(3 ones) as:

$$P(3\,\text{ones}) = \binom{10}{3}p^3(1-p)^7$$

Therefore, the probability of an event with probability p, occurring k times in n trials can be generalized to:

[48] $\quad P_n(k) = \dbinom{n}{k}p^k(1-p)^{n-k}$

Example 2.7:

Let us consider another application. Say we look at 4 bits from the output of a digital circuit. If the probability of a 1 is p, what is the probability of seeing at least three 1s. This would be found by adding the probability of seeing three 1s and the probability of seeing four 1s as follows:

$$\binom{4}{3}p^3(1-p)^1 + \binom{4}{4}p^4(1-p)^0 \quad \blacksquare$$

One last thing to look at before finishing this chapter is how to evaluate [48] when n is large. This is resolved using the DeMoivre-Laplace theorem which is stated as:

[49] $\quad P_n(k) = \dbinom{n}{k}p^k(1-p)^{n-k} \approx \dfrac{e^{-\frac{(k-np)^2}{2np(1-p)}}}{\sqrt{2\pi np(1-p)}}$

To use this theorem the following two conditions must be satisfied:

- $np(1-p) >> 1$

- $|k-np|$ must be on the same order as or less than $(np(1-p))^{1/2}$

2.9   Exercises

1 Due to noise, the transmitted data in a binary communication channel occasionally gets corrupted; a 1 gets turned into a 0 and a 0 is turned into a 1. The

probability of transmitting a 1 is P(X=1)=0.7 and the probability of transmitting a 0 is P(X=0)=1-0.7=0.3. Also, the probability that a transmitted 0 is received as a 1 is P(Y=1|X=0)= 0.01 and the probability that a transmitted 1 is received as a 0 is P(Y=0|X=1)= 0.05.

   a)   What is the probability for transmitting a 1 correctly and what is the probability of transmitting a 0 correctly?

   b)   What is the probability for receiving a 1 and what is the probability of receiving a 0?

   c)   Determine the four conditional probabilities P(X=0|Y=0), P(X=0|Y=1), P(X=1|Y=0), and P(X=1|Y=1).

   d)   What is the probability that any bit would be in error?

2   Five defective devices are accidently dumped into a bin containing 995 known good devices. There fortunately is a test that can be performed which can tell if a device is good or defective 95% of the time. A device is picked from the bin, tested, and found to be good. What is the probability that the device is actually good?

3   Determine P(0.7<V<0.85|V>0.4) for the triangular waveform below.

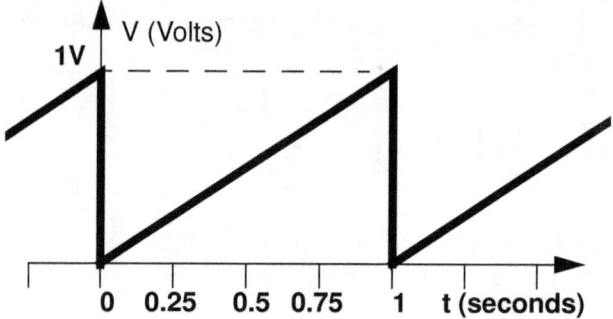

Figure 27   Figure for exercise 3

4   P(A)=0.2, P(B)=0.3, and P(C)=0.1. If these three events are independent, list all of the possible joint probabilities that must exist.

5   Consider two mutually exclusive events A and B (i.e. A ∩ B = ∅) where P(A)≠0 and P(B)≠0. Are these events independent? What general conclusion could you draw from the result?

6   Consider [48]. Write the summations for the following:

   a)   P(A occurs less than k times in n trials)

   b)   P(A occurs more than k times in n trials)

c) P(A occurs no more than k times in n trials)

d) P(A occurs at least k times in n trials)

7 The probability for switches A - D to be closed are 0.1, 0.05, 0.85, and 0.7 respectively. What is the probability that there exists a path from point 1 to point 2?

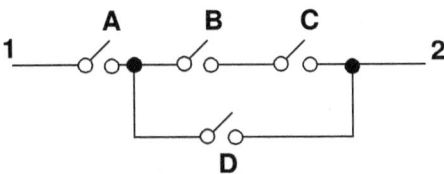

Figure 28 Switch circuit for exercise 7

8 A transmitted block of binary data consists of 5-bits. The probability of a bit becoming errored is p=0.01. Find the following probabilities:

a) No bits are errored.

b) Less than 2-bits are errored.

9 Determine $P(0.7 < V < 0.85 | V > 0.6)$ for the periodic waveform below.

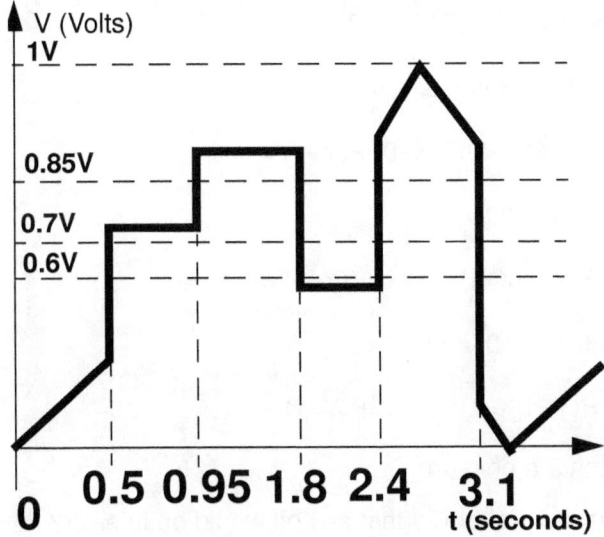

Figure 29 Figure for exercise 9

2.10   Solutions to Exercises

1

33

a) What is the probability for transmitting a 1 correctly and what is the probability of transmitting a 0 correctly?

$P(Y=1|X=1) = 1 - P(Y=0|X=1) = 1 - 0.05 = 0.95$

$P(Y=0|X=0) = 1 - P(Y=1|X=0) = 1 - 0.01 = 0.99$

b) What is the probability for receiving a 1 and what is the probability of receiving a 0?

$P(Y=1) = P(Y=1|X=1)P(X=1) + P(Y=1|X=0)P(X=0) =$

$(0.95)(0.7) + (0.01)(0.3) = 0.668$

$P(Y=0) = P(Y=0|X=1)P(X=1) + P(Y=0|X=0)P(X=0) =$

$(0.05)(0.7) + (0.99)(0.3) = 0.332$

Note that $P(Y=1) + P(Y=0) = 1$ as would be expected.

c) Determine the four conditional probabilities $P(X=0|Y=0)$, $P(X=0|Y=1)$, $P(X=1|Y=0)$, and $P(X=1|Y=1)$.

$P(X=0|Y=0) = P(Y=0|X=0)P(X=0)/P(Y=0) =$

$(0.99)(0.3)/0.332 = 0.894578$

$P(X=0|Y=1) = P(Y=1|X=0)P(X=0)/P(Y=1) =$

$(0.01)(0.3)/0.668 = 0.004491$

$P(X=1|Y=0) = P(Y=0|X=1)P(X=1)/P(Y=0) =$

$(0.05)(0.7)/0.332 = 0.105422$

$P(X=1|Y=1) = P(Y=1|X=1)P(X=1)/P(Y=1) =$

$(0.95)(0.7)/0.668 = 0.995509$

d) What is the probability that any bit would be in error?

$P(error) = P(Y=0|X=1)P(X=1) + P(Y=1|X=0)P(X=0) =$

$(0.05)(0.7) + (0.01)(0.3) = 0.038$

2 We can use Bayes' theorem. Let's define A = {tested device is not defective} and B = {test result indicates that device is not defective}.

$$P(A|B) = \frac{P(B|A)P(A)}{P(B|A)P(A) + P(B|A^C)P(A^C)}$$

$P(A) = 995/(995+5) = 0.995$

$P(A^C) = 1 - 0.995 = 0.005$

$P(B|A) = 0.95$

$P(B|A^C) = 1 - 0.95 = 0.05$

Thus, $P(A|B) = [(0.95)(0.995)] / [(0.95)(0.995) + (0.05)(0.005)] = 0.999736$

3 Since the waveform is periodic we only need to consider one cycle. $P(0.7<V<0.85|V>0.4)$ is equal to:

$P(0.7<V<0.85 \cap V>0.4) / P(V>4)$

Referring to Figure 30:

The number of elements in Ω is equal to the total length of time being considered which is 1-0=1second.

The number of elements in $0.7<V<0.85 \cap V>0.4$ is equal to the length of time that that the waveform is between 0.7 volts and 0.8 volts while also being above 0.4 volts[1]. This is equal to 0.85-0.7= 0.15 seconds.

The number of elements in $V>0.4$ is the length of time that the waveform is above 0.4 volts. This is equal to 1-0.4=0.6 seconds.

$P(0.7<V<0.85 \cap V>0.4)$ is equal to:

$$\frac{\text{\# of elements in } 0.7<V<0.85 \cap V>0.4}{\text{\# of elements in } \Omega} = 0.15/1$$

$$\frac{\text{\# of elements in } V>0.4}{\text{\# of elements in } \Omega} = 0.6/1$$

---

1. While it may sound redundant to state that 0.7volts-0.85volts is above 0.4volts this may not always be the case. For example if the conditioning event was changed to V>0.75volts then the part of the waveform from 0.7volts to 0.75volts would not be included in the calculation.

Therefore, 0.15/0.6 = 0.4 is the answer.

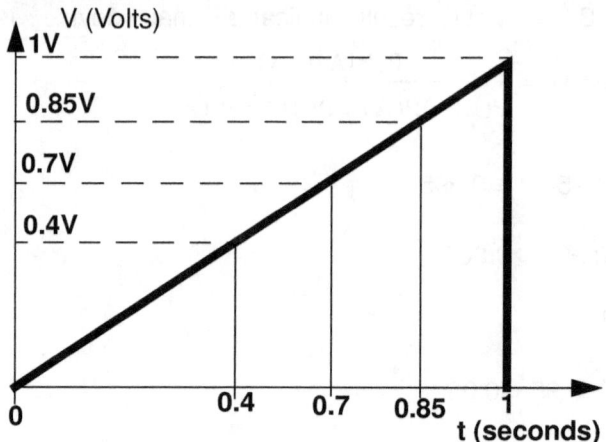

Figure 30 Figure for exercise 3

4   To ensure independence, all combinations of the three events must be shown to be independent.

P(A ∩ B) = P(A)P(B) = 0.06

P(A ∩ C) = P(A)P(C) = 0.02

P(B ∩ C) = P(B)P(C) = 0.03

P(A ∩ B ∩ C) = P(A)P(B)P(C) = 0.006

5 For independence, P(A ∩ B)=P(A)P(B). If A nd B are mutually exclusive events, then P(A ∩ B)=P(∅)=0. This would imply that either P(A) or P(B) or both must be 0, which they are not. Therefore, this implies that they are not independent. The conclusion that we can draw from this is that just because events are disjoint does not necessarily mean that they are independent.

6

   a) P(A occurs less than k times in n trials)

The probability we seek consists of the sum of the $P_n(0)$, $P_n(1)$, $P_n(2)$,...$P_n(k-2)$, $P_n(k-1)$. Thus, using a summation:

$$P(\text{A occurs} < k \text{ times in n trials}) = \sum_{i=0}^{k-1} P_n(i)$$

b) P(A occurs more than k times in n trials)

The probability we seek consists of the sum of the $P_n(k+1)$, $P_n(k+2)$,...$P_n(n-1)$, $P_n(n)$. Thus, using a summation:

$$P(\text{A occurs} > k \text{ times in n trials}) = \sum_{i=k+1}^{n} P_n(i)$$

c) P(A occurs no more than k times in n trials)

The probability we seek consists of the sum of the $P_n(0)$, $P_n(1)$, $P_n(2)$,...$P_n(k-1)$, $P_n(k)$. Thus, using a summation:

$$P(\text{A occurs} \leq k \text{ times in n trials}) = \sum_{i=0}^{k} P_n(i)$$

d) P(A occurs at least k times in n trials)

The probability we seek consists of the sum of the $P_n(k)$, $P_n(k+1)$,...$P_n(n-1)$, $P_n(n)$. Thus, using a summation:

$$P(\text{A occurs} \geq k \text{ times in n trials}) = \sum_{i=k}^{n} P_n(i)$$

7 Because the state of each switch does not affect the states of the other switches we can assume that they open and close independently. In order for a path to exist between points 1 and 2 switch A must be closed and either switches B and C, or switch D must be closed. Thus we can write P(path from 1 to 2) = P[A ∩ ((B ∩C) ∪D)]. Because of independence and [45] this becomes:

$P[A \cap ((B \cap C) \cup D)] = P(A)P((B \cap C) \cup D) = P(A)[P(B)P(C) + P(D) - P(B)P(C)P(D)] = (0.1)[(0.05)(0.85) + 0.7 - (0.05)(0.85)(0.7)] = 0.0445.$

8

a) No bits are errored. Using [48]:

$$P_5(0) = \binom{5}{0}(0.01)^0(1 - 0.01)^{5-0}$$

This evaluates to 0.951.

b) Less than 2-bits are errored. This corresponds to $P_5(0) + P_5(1)$. Using [48] again:

$$P_5(0) + P_5(1) = 0.951 + \binom{5}{1}(0.01)^1(1 - 0.01)^{5-1}$$

This evaluates to 0.999.

9 This exercise is handled in the same way as was exercise 3.

The length of time that V is greater than 0.6 volts is (3.1-2.4) + (1.8-0.5) = 2 seconds.

The length of time that V is between 0.7 volts and 0.85 volts while also being above 0.6 volts is 0.95 - 0.5 = 0.45 seconds.

Therefore, $P(0.7<V<0.85|V>0.6) = 0.45/2 = 0.225$.

Chapter 3   Midterm Quiz

The questions below cover the material presented in chapters 1 and 2. To get the most benefit from these questions try working them out first before looking at the solutions or the previous chapters. If you absolutely cannot solve a problem, then try looking back over the material in the previous chapters before looking at the solutions (which should be the last resort). These questions are pretty straight forward and should not take more than an hour and a half to complete. Calculators are allowed.

3.1   Midterm Quiz

Q.1 Draw the Venn diagrams for the following situations given the events A, B, and C:

a Exactly two events occur.

b If A or B occur then C does not occur.

c Show $(A - B) \cap (A - C)$.

Q.2 Given that A, B, and C are events, and are not disjoint, find $P(A \cup B \cup C)$.

Q.3 Prove the following two relationships:

a $A \cap (B \cup C) = (A \cap B) \cup (A \cap C)$

b $A \cup (B \cap C) = (A \cup B) \cap (A \cup C)$

Q.4 A bin contains 50 type A components, 35 type B components, and 20 type C components. 10% of the type A components are defective and 25% of the type C components are defective.

a If a device is selected at random, what is the probability that it is defective?

b If a device is selected at random, what is the probability that it is type A and not defective?

c If a device is selected at random and it is found to be defective, what is the probability that it is a type C device?

d If a device is selected and it is not defective, what is the probability that it is a type B device?

Q.5 Derive a general expression that gives the number of equations needed to prove independence. Hint: use [47] and the fact that $\sum_{k=0}^{n} \binom{n}{k} = 2^n$.

Q.6 In a certain communications link the probability of transmitting a 0 is 0.3 and the probability of transmitting a 1 is 0.7. Because of noise, the bits sometimes get corrupted and, therefore, a transmitted 1 is corrupted with probability 0.08 and a transmitted 0 is corrupted with probability of 0.02. Therefore:

a What is the probability of receiving 7 consecutive bits with exactly one error?

b What is the probability of receiving 7 consecutive bits with 3 or more errors?

Q.7 An experiment has three disjoint outcomes A_1, A_2, and A_3 and another experiment has two disjoint outcomes B_1 and B_2. The joint probabilities for these outcomes are: $P(A_1 \cap B_1) = 0.1$, $P(A_1 \cap B_2) = 0.1$, $P(A_2 \cap B_1) = 0.1$, $P(A_2 \cap B_2) = 0.2$, $P(A_3 \cap B_1) = 0.2$, $P(A_3 \cap B_2) = 0.3$.

a Are all of these outcomes independent?

b What are all of the conditional probabilities?

Q.8 A bin of resistors contains 50 10Ω resistors, 55 20Ω resistors, and 100 40Ω resistors. A resistor is drawn at random.

a What is the probability that the first resistor drawn is 10Ω?

b If the first resistor drawn is 10Ω and is not replaced, what is the probability that the 2nd resistor drawn is also 10Ω?

Q.9 In a certain digital device the probability for any bit to be in error is p. There is a function in this device that looks at a group of three bits and uses a majority vote to determine if a one or a zero has been received[1]. What is the probability that this determination is incorrect?

---

1. e.g., 111, 110, 101, or 011 would be interpreted as a 1. 000, 001, 010, or 100 would be interpreted as a 0.

3.2   Solutions to Midterm Quiz

A.1 Hint: for a and b it will be easier if we 1st convert the description into symbols which can then be converted into a Venn diagram.

a Exactly two events occur means $(A \cap B \cap C^C) \cup (A \cap B^C \cap C) \cup (A^C \cap B \cap C)$. It may be less confusing if we 1st draw the three intersections separately and then take their union as shown below.

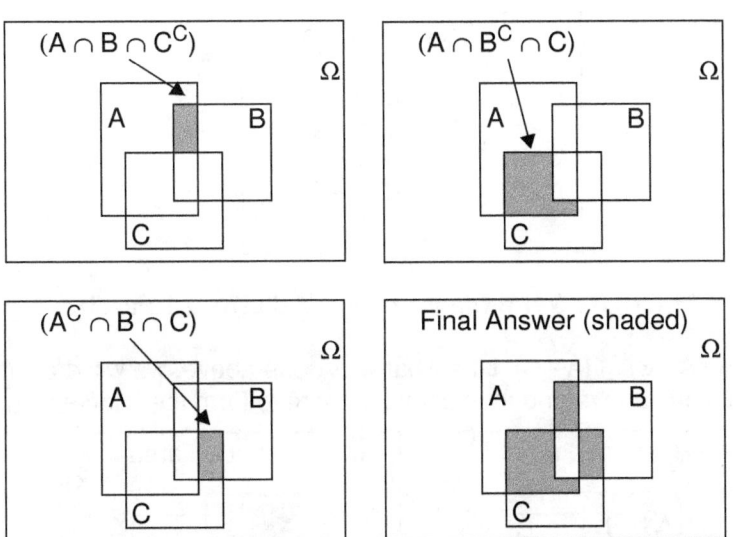

Figure 31 Venn Diagrams for Midterm Q.1 Part a

b If A or B occur then C does not occur means $(A \cup B) \cap C^C$. It may be less confusing if we 1st draw $(A \cup B)$ and C^C separately and then combine them to create their intersection.

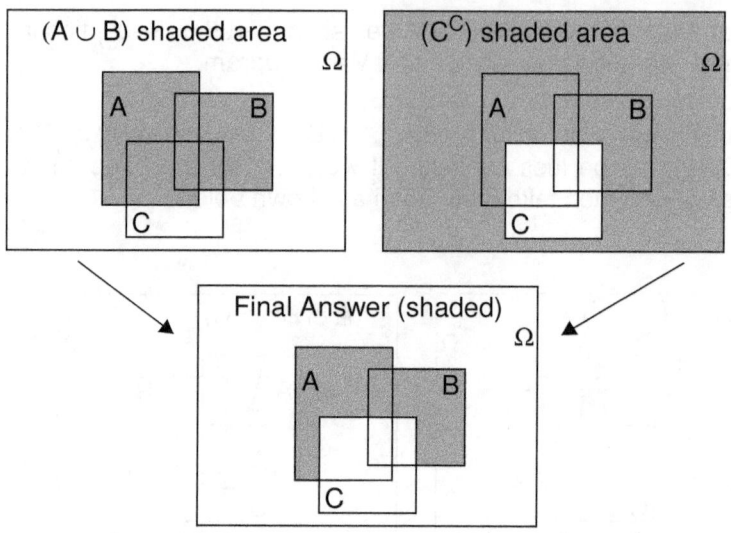

Figure 32 Venn Diagrams for Midterm Q.1 Part b

c Show (A - B) ∩ (A - C). Like what was done above, we will draw the two subtractions above and then combine them to form the intersection.

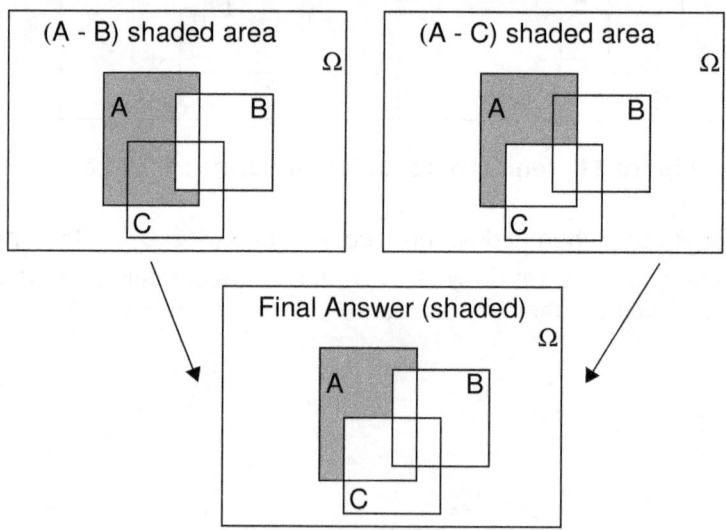

Figure 33 Venn Diagrams for Midterm Q.1 Part c

A.2 We can come to the answer by repeated application of [28].

[50] $P(A \cup B \cup C) = P(A \cup (B \cup C)) = P(A) + P(B \cup C) - P(A \cap (B \cup C))$.

[51] $P(B \cup C) = P(B) + P(C) - P(B \cap C)$.

Using [15] we have:

[52] $P(A \cap (B \cup C)) = P((A \cap B) \cup (A \cap C)) = P(A \cap B) + P(A \cap C) - P(A \cap B \cap A \cap C) = P(A \cap B) + P(A \cap C) - P(A \cap B \cap C)$.

Plugging [51] and [52] into [50] provides the answer:

$P(A \cup B \cup C) = P(A) + P(B) + P(C) - P(A \cap B) - P(A \cap C) - P(B \cap C) + P(A \cap B \cap C)$.

A.3 To prove the relationships, a Venn Diagram is drawn for each side of the equation and, when compared, both Venn Diagrams are verified to be equal.

a $A \cap (B \cup C) = (A \cap B) \cup (A \cap C)$

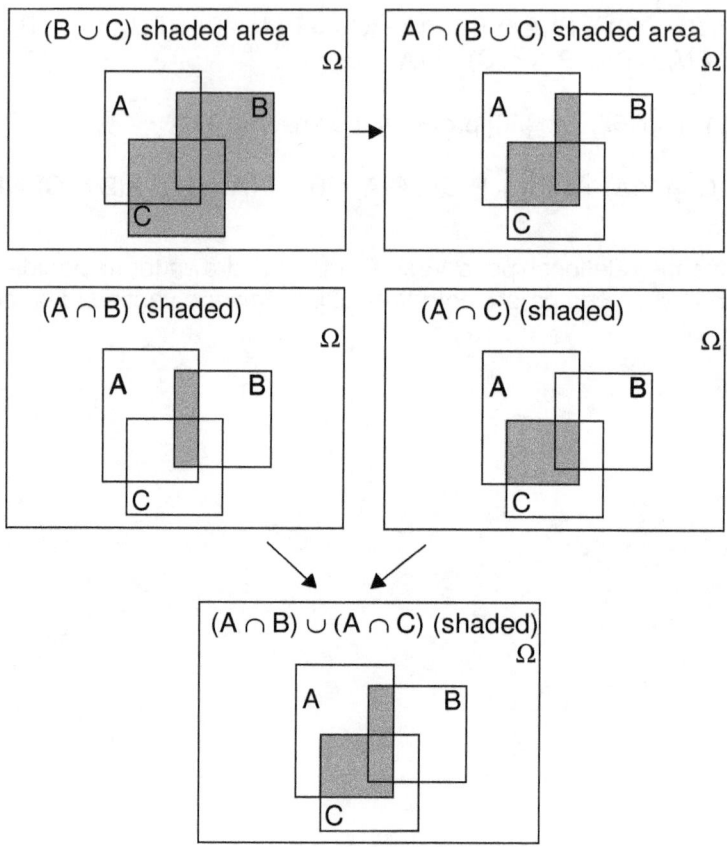

Figure 34 Venn Diagrams for Midterm Q.3 Part a

b $A \cup (B \cap C) = (A \cup B) \cap (A \cup C)$

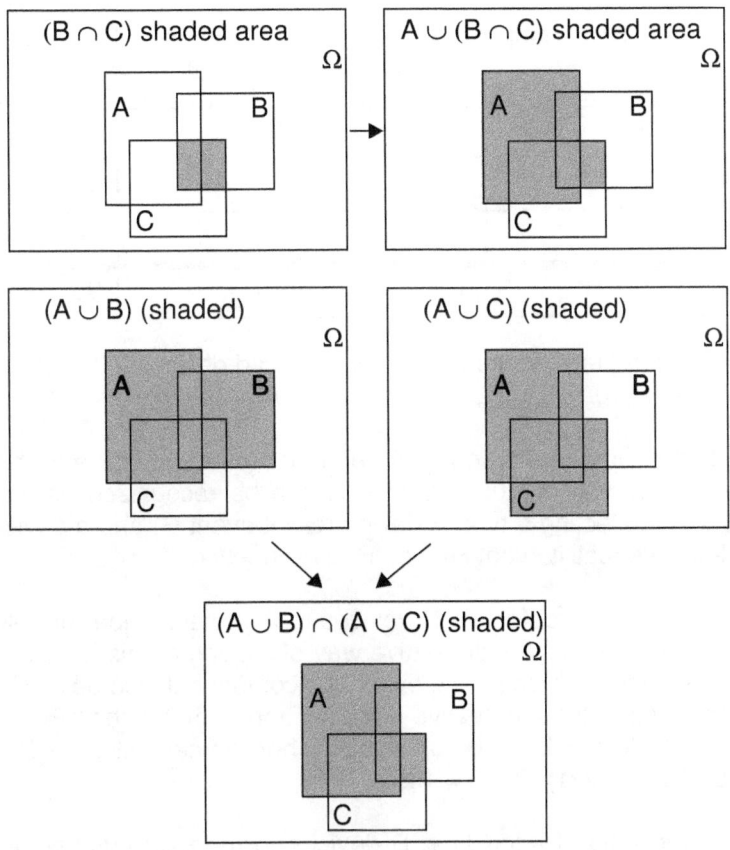

Figure 35 Venn Diagrams for Midterm Q.3 Part b

A.4 Let's start out by tabulating the values:

Table 1 Q.4

	Component Type	Total Quantity	Quantity Defective	Quantity Good
	A	50	5	45
	B	35	0	35
	C	20	5	15
Totals		105	10	95

a This is just the ratio of the total number of bad devices to the total number of devices: $10/105 = 2/21$.

b This is the number of type A devices that are not defective to the total number of devices: $45/105 = 3/7$. This can be recognized as the joint probability of selecting a type A device (call it event G) and a device that is not defective (call it event H). i.e., $P(G \cap H) = 3/7$.

c This is the ratio of defective C devices to the total number of defective devices: $= 5/10 = 1/2$. An alternative way of looking at this is as a conditional probability. Let's call K the event of a component type being C and L the event of selecting a defective device. Then from the chart $P(K \cap L) = 5/105 = 1/21$. Also $P(L) = 10/105 = 2/21$. Then we can calculate $P(K|L) = P(K \cap L)/P(L) = (1/21)/(2/21) = 1/2$.

d This is the ratio of good type B devices to the total number of good devices: $35/95 = 7/19$. This can also be solved using conditional probabilities. Let's call event V the event of selecting a good part. From the chart $P(V) = 95/105 = 19/21$. Let's call event W the event that a type B part is selected. From the chart $P(V \cap W) = 35/105 = 7/21$. Therefore, $P(W|V) = P(V \cap W)/P(V) = (7/21)/(19/21) = 7/19$.

A.5 To determine independence, [42]-[44] must be satisfied. This means that independence must be shown for all pairs, triples, quadruples etc. The binomial coefficient provides us with how many groupings of k objects can be made from n objects.

The number of pairs that can be made is $\binom{n}{2}$. e.g., $P(A_1)P(A_2)$, $P(A_1)P(A_3)$,...etc.

The number of triples that can be made is $\binom{n}{3}$. $P(A_1)P(A_2)P(A_3)$, $P(A_1)P(A_2)P(A_4)$,...etc.

The number of quadruples that can be made is $\binom{n}{4}$, etc., etc., all the way up

to $\binom{n}{n}$. This can be written as $\sum\limits_{k=2}^{n} \binom{n}{k}$. We can represent this summation as

$$\sum_{k=2}^{n} \binom{n}{k} = \sum_{k=0}^{n} \binom{n}{k} - \binom{n}{0} - \binom{n}{1}.$$ Replacing $\sum\limits_{k=0}^{n} \binom{n}{k}$ with 2^n, and evaluating

the binomial coefficients we get: $2^n - 1 - n$.

A.6 We first find the probability that any bit would be in error. This is $p = (0.3)(0.02) + (0.7)(0.08) = 0.062$.

a For exactly one bit in error: $p^1(1-p)^6$. Now we need to multiply this by the number of ways that 1 bit can be errored in a set of seven bits. This is found by the binomial coefficient. Therefore, the answer is:

$$\binom{7}{1}p^1(1-p)^6 = 0.296$$

b In this case we nee to sum up the cases where the errors are 3 to 7.

Which results in: $\sum\limits_{k=3}^{7} \binom{7}{k}p^k(1-p)^{7-k} = 6.901(10^{-3})$

A.7 In this problem all of the joint probabilities are disjoint from each other and add up to 1 so we have a partition of the sample space as shown below.

$A_1 \cap B_1$	$A_2 \cap B_1$	$A_3 \cap B_1$
$A_1 \cap B_2$	$A_2 \cap B_2$	$A_3 \cap B_2$

Figure 36 Venn Diagram for Midterm Question Q.7

a We know from Chapter 2 exercise 5, that these disjoint events cannot be independent so events A_1-A_3 are not independent; the same goes for B_1 and B_2. We can however solve for the marginal probabilities using the total probability theorem and see if any of the individual combinations of events are independent.

$P(A_1) = P(A_1 \cap B_1) + P(A_1 \cap B_2) = 0.1 + 0.1 = 0.2$.

$P(A_2) = P(A_2 \cap B_1) + P(A_2 \cap B_2) = 0.1 + 0.2 = 0.3$.

$P(A_3) = P(A_3 \cap B_1) + P(A_3 \cap B_2) = 0.2 + 0.3 = 0.5$.

$P(B_1) = P(A_1 \cap B_1) + P(A_2 \cap B_1) + P(A_3 \cap B_1) = 0.1 + 0.1 + 0.2 = 0.4$.

$P(B_2) = P(A_1 \cap B_2) + P(A_2 \cap B_2) + P(A_3 \cap B_2) = 0.1 + 0.2 + 0.3 = 0.6$.

$P(A_1 \cap B_1) = 0.1 \neq P(A_1)P(B_1) = (0.2)(0.4) = 0.08$

$P(A_1 \cap B_2) = 0.1 \neq P(A_1)P(B_2) = (0.2)(0.6) = 0.12$

$P(A_2 \cap B_1) = 0.1 \neq P(A_2)P(B_1) = (0.3)(0.4) = 0.12$

$P(A_2 \cap B_2) = 0.2 \neq P(A_2)P(B_2) = (0.3)(0.6) = 0.18$

$P(A_3 \cap B_1) = 0.2 = P(A_3)P(B_1) = (0.5)(0.4) = 0.2$

$P(A_3 \cap B_2) = 0.3 = P(A_3)P(B_2) = (0.5)(0.6) = 0.3$

Therefore, we can conclude that events A_3 and B_1 are independent and events A_3 and B_2 are independent.

b The conditional probabilities are:

$P(A_1|B_1) = P(A_1 \cap B_1)/P(B_1) = 0.1/0.4 = 1/4$

$P(A_1|B_2) = P(A_1 \cap B_2)/P(B_2) = 0.1/0.6 = 1/6$

$P(A_2|B_1) = P(A_2 \cap B_1)/P(B_1) = 0.1/0.4 = 1/4$

$P(A_2|B_2) = P(A_2 \cap B_2)/P(B_2) = 0.2/0.6 = 1/3$

$P(A_3|B_1) = P(A_3 \cap B_1)/P(B_1) = 0.2/0.4 = 1/2$

$P(A_3|B_2) = P(A_3 \cap B_2)/P(B_2) = 0.3/0.6 = 1/2$

$P(B_1|A_1) = P(A_1 \cap B_1)/P(A_1) = 0.1/0.2 = 1/2$

$P(B_2|A_1) = P(A_1 \cap B_2)/P(A_1) = 0.1/0.2 = 1/2$

$P(B_1|A_2) = P(A_2 \cap B_1)/P(A_2) = 0.1/0.3 = 1/3$

$P(B_2|A_2) = P(A_2 \cap B_2)/P(A_2) = 0.2/0.3 = 2/3$

$P(B_1|A_3) = P(A_3 \cap B_1)/P(A_3) = 0.2/0.5 = 2/5$

$P(B_2|A_3) = P(A_3 \cap B_2)/P(A_3) = 0.3/0.5 = 3/5$

A.8 This is a simple problem.

a This is simply the ratio of 10Ω resistors to the total number of resistors: $50/(50+55+100) = 50/205 = 10/41$.

b The total number of 10Ω resistors left is 49 and the total number of resistors left is 204, therefore, the probability is $49/204$.

A.9 For an error to occur two or three bits must be found in error. For two bits to be in error we can write:

$\binom{3}{2}p^2(1-p)^{3-2}$. This gives the probability that two out of the three bits can be in error. For three bits in error we can write:

$\binom{3}{3}p^3(1-p)^{3-3}$. These two pieces can be summed together as:

$$P(error) = \sum_{k=2}^{3} \binom{3}{k}p^k(1-p)^{3-k}$$ to yield $3p^2(1-p) + p^3$. In most systems p is much less than 1, thus this method makes it less likely to incorrectly decode the intended value, which makes for a more robust design.

Chapter 4   Random Variables

In the previous chapters we looked at determining sample spaces, defining events and assigning probabilities to those events. In this chapter, we will discuss what random variables are and develop some of the tools to work with them.

4.1   Definition of a Random Variable

A random variable is simply the mapping of each of the elements (ζ) of the sample space Ω into real numbers. Random variables will be denoted by capital letters with the element as the independent variable, e.g., $X(\zeta)$.

Example 4.1:

Consider the binary sequence Ω = {000, 001, 010, 011, 100, 101, 110, 111}. A possible mapping could be:

$X(000) \to 0$

$X(001) \to 1$

$X(010) \to 2$

$X(011) \to 3$

$X(100) \to 4$

$X(101) \to 5$

$X(110) \to 6$

$X(111) \to 7$ ∎

Example 4.2:

Let's consider an example where the sample space is continuous. The phase of the vector in the figure below varies randomly as shown from θ_1 to θ_2. The assignment is made as $X(\theta_1) = \theta_1 \leq X(\theta) \leq X(\theta_2) = \theta_2$. Note that any numbers could have been used, but mapping directly is simpler.

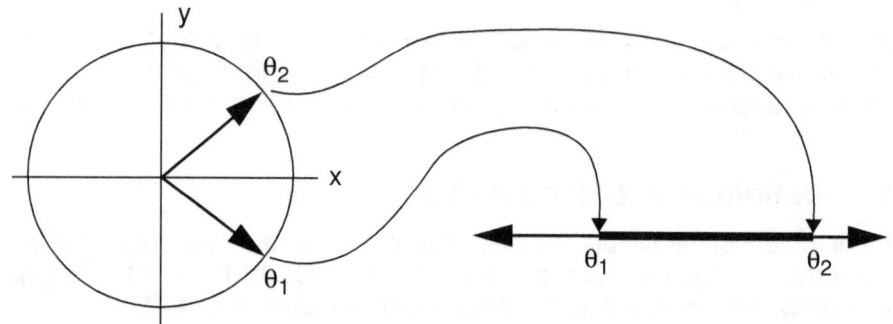

Figure 37 Mapping of a Continuous Sample Space to the Real Line

■

When working with random variables, it is customary to drop the independent variable notation so that X = X(ζ). It should also be noted that the mapping of the sample space to real numbers is essentially arbitrary.

4.2   Assigning Probabilities to Random Variables

The typical probabilities that are sought for random variables are $P(X \leq x)$ or $P(x_1 \leq X \leq x_2)$ for the continuous case and $P(X=x)$ for the discrete case. The set $\{X \leq x\}$ consists of all of the elements (ζ) of Ω for which $X(\zeta) \leq x$. If $\Omega = \{00, 01, 10, 11\}$ and X(00)=1, X(01)=2, X(10)=3, X(11)=4, then set $\{X \leq 2.5\} = \{00, 01\}$. Notice that it is not equal to $\{1, 2\}$ but instead contains the actual elements of Ω. Now we can find $P(X \leq 2.5)$ as 2/4=0.5. The same principle applies for finding $P(x_1 \leq X \leq x_2)$. We will now consider a few examples.

Example 4.3:

Consider the following function $x(t)=\sin(\omega t + \theta)$, where θ varies randomly and uniformly from 0 to 2π.   What is $P(0.1\pi \leq \Theta \leq 1.3\pi)$? First Ω is $2\pi-0=2\pi$. Next we have $\{0.1\pi \leq \Theta \leq 1.3\pi\}$ as the event of interest. Therefore, $P(0.1\pi \leq \Theta \leq 1.3\pi) = (1.3\pi-0.1\pi)/2\pi = 0.6$.■

Example 4.4:

Consider 3-bits observed at random $\Omega = \{000, 001, 010, 011, 100, 101, 110, 111\}$. The probability of each element is uniformly distributed so the probability of each element is 0.125. Furthermore, let us generate the random variables as X(000)=0 - X(111)=7. Therefore:

$P(X \leq 0)=P(\{000\})=1/8$

$P(X \le 1)=P(\{000, 001\})=2/8=1/4$

$P(X \le 2)=P(\{000, 001, 010\})=3/8$

$P(X \le 3)=P(\{000, 001, 010, 011\})=4/8=1/2$

$P(X \le 4)=P(\{000, 001, 010, 011, 100\})=5/8$

$P(X \le 5)=P(\{000, 001, 010, 011, 100, 101\})=6/8=3/4$

$P(X \le 6)=P(\{000, 001, 010, 011, 100, 101, 110\})=7/8$

$P(X \le 7)=P(\{000, 001, 010, 011, 100, 101, 110, 110\})=8/8=1$ ■

In the two examples above the probabilities of each element in Ω are equal. Let's consider a case where the probabilities are not equal.

Example 4.5:

Consider again the 3-bits observed at random $\Omega = \{000, 001, 010, 011, 100, 101, 110, 111\}$. Suppose now that we are interested in the number of 1s in each 3-bit sample. The mapping of the elements of the sample space can be done as follows:

$X(000) \rightarrow 0$

$X(001) \rightarrow 1$

$X(010) \rightarrow 1$

$X(011) \rightarrow 2$

$X(100) \rightarrow 1$

$X(101) \rightarrow 2$

$X(110) \rightarrow 2$

$X(111) \rightarrow 3$

It is allowable to map multiple outcomes to a single number, however mapping a single outcome to multiple numbers is not allowed. Now the probabilities can be defined as:

$P(X=0) = P(\{000\}) = 1/8$

$P(X=1) = P(\{001,010,100\}) = 3/8$

$P(X=2) = P(\{011,101,110\}) = 3/8$

$P(X=3) = P(\{111\}) = 1/8$ ∎

4.3 Cumulative Distribution Functions (cdf)

The cumulative distribution function is defined as $F_X(x) = P(X \leq x)$. Throughout this book $F_X(X \leq x)$ is sometimes used interchangeably with $F_X(x)$.

Example 4.6:

What is the cdf for example 4.5 above? Since the probabilities only exist for discrete values it only makes sense to use discrete values for x. Therefore:

$F_X(x) = 0$ for x less than 0 since we cannot have less than zero 1s in any 3-bit sample.

$F_X(0) = P(X \leq 0) = P(X=0) = 1/8$

$F_X(1) = P(X \leq 1) = P(X=0) + P(X=1) = 1/8 + 3/8 = 1/2$

$F_X(2) = P(X \leq 2) = P(X=0) + P(X=1) + P(X=2) = 1/8 + 3/8 + 3/8 = 7/8$

$F_X(3) = P(X \leq 3) = P(X=0) + P(X=1) + P(X=2) + P(X=3) = 1/8 + 3/8 + 3/8 + 1/8 = 1$. For "x" greater than 3 $F_X(x)$ will always be a 1 since the number of 1s will never exceed 3 in any 3-bit sample.

We can see that as "x" gets larger, the probabilities accumulate, hence the name cdf. Below is a plot of the cdf:

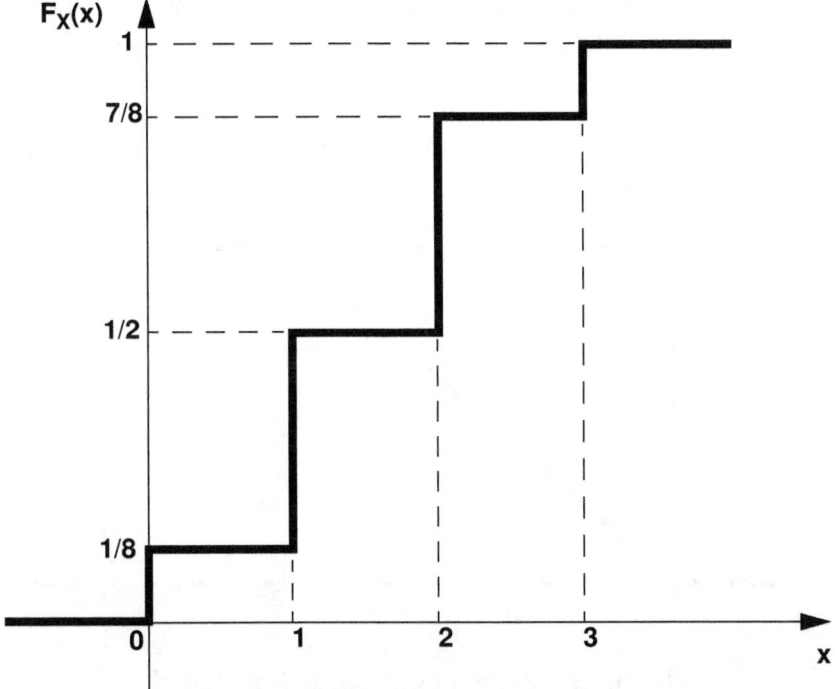

Figure 38 cdf for example 4.5

Finally, we could represent this cdf in closed form as a sum of step functions:

$$[53] \quad P(X \le x) = \sum_{i=0}^{3} p_i U(x-i)$$

where U(x-i) is the unit step function[1] and p_0 =1/8, p_1 =3/8, p_2 =3/8, and p_3 =1/8 ■

Example 4.7:

Let's look at an example for a continuous case. Consider again the random variable θ from example 4.3. $F_\Theta(\theta) = P(\Theta \le \theta)$ is easily seen

---

1. U(x-i) = 1 for x≥i and is 0 for x<i.

to be $\theta/(2\pi-0) = \theta/(2\pi)$ for $0 \leq \theta \leq 2\pi$. Since θ never goes above 2π $F_\Theta(\theta) = 1$ for $\theta \geq 2\pi$ and $F_\Theta(\theta) = 0$ for $\theta < 0$ since θ never goes below 0. The plot of this cdf is:

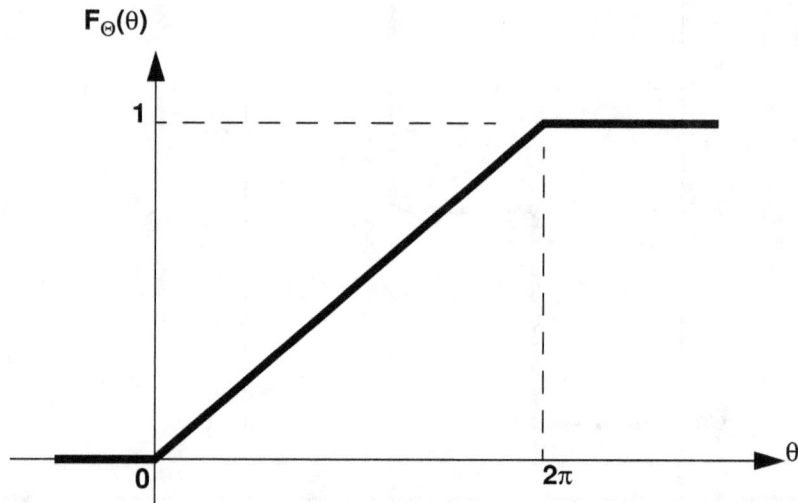

Figure 39 cdf for a continuous function

■

There are several useful properties of cdfs. These are:

[54] $F_X(\infty)= P(X \leq \infty) = 1$ (this is intuitive from the two examples above).

[55] $F_X(-\infty)= P(X \leq -\infty) = 0$ (this also is intuitive from the two examples above).

[56] $F_X(x_2) - F_X(x_1) = P(x_1 < X \leq x_2), \quad x_1 < x_2$

[57] $1 - F_X(x) = P(X > x)$

Also notice that $F_X(x)$ increases as "x" increases. It never decreases as "x" increases. Thus we say that $F_X(x)$ monotonically increases. This is because of the cumulative nature of the cdf function.

4.4   Probability Density Functions

The probability density function (pdf) is an alternative way of looking at the cdf and is defined as the derivative of the cdf function:

[58] $f_X(x) = \dfrac{d}{dx}F_X(x)$

There are several useful properties of the pdf as follows:

[59] $\displaystyle\int_{-\infty}^{\infty} f_X(\xi)d\xi = 1$ [1]

[60] $\displaystyle\int_{-\infty}^{x} f_X(\xi)d\xi = P(X \leq x) = F_X(x)$

[61] $\displaystyle\int_{x_1}^{x_2} f_X(\xi)d\xi = P(x_1 < X \leq x_2)$

[62] $f_X(x) \geq 0$ for all x

Example 4.8:

Consider again the random variable θ from example 4.3 and the plot of it's cdf in Figure 39. Its pdf is the derivative of the cdf, which is shown below.

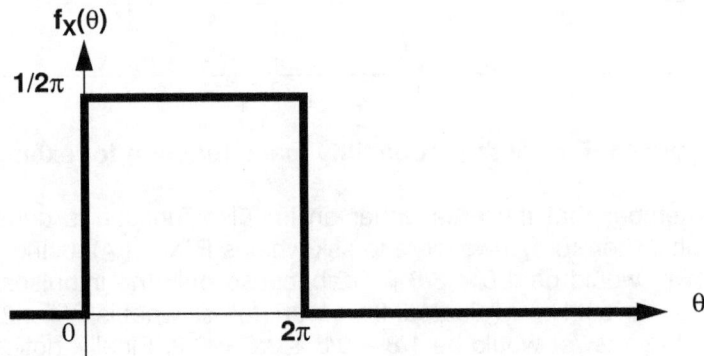

Figure 40 Uniform Probability Density Function

---

1. The interpretation of this is that the area under the pdf is 1.

Notice that its area is 1 per [59]. If we were now to ask what is the probability that θ lies in the range from 0.1 to 0.3 we would calculate it as follows:

$$P(0.1 \le \theta \le 0.3) = \int_{0.1}^{0.3} \frac{1}{2\pi} d\xi = \frac{1}{2\pi}\bigg|_{0.1}^{0.3} = \frac{1}{10\pi} \quad \blacksquare$$

For a discrete random variable there is a function that is sometimes referred to as a probability mass function. It is defined for countable events and is defined as the probability that a random variable is exactly equal to a number.

Example 4.9:

Consider the cdf in example 4.6 above. Taking its derivative, and using impulses[1] to represent the derivative at the discontinuities, we have the probability mass plot as shown below.

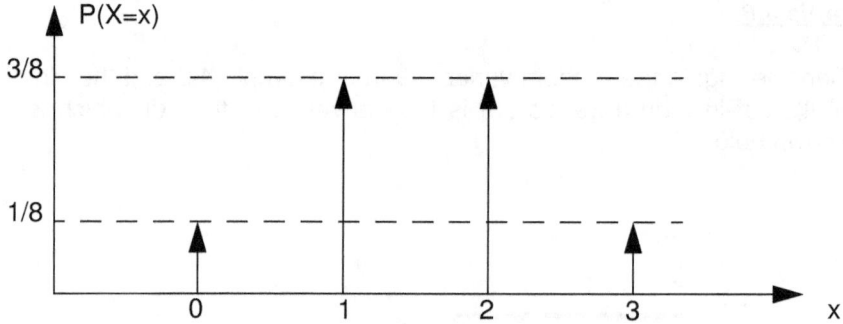

Figure 41 Plot of the probability mass function for example 4.6

Remember that the area under an impulse function is equal to its height. Therefore, if we were to ask what is $P(X \le 1.4)$, using [60] the answer would be 1/8 + 3/8 = 1/2 because only the impulses at x=0 and x=1 would be evaluated. If we were to ask what is $P(X \le 2)$, using [60] the answer would be 1/8 + 3/8 + 3/8 = 7/8. Finally, notice that if [59] was used then the area of $f_X(x)$ is 1 as expected. ■

---

1. An impulse is represented by δ(argument). Whenever the argument is 0, the impulse takes on a value of 1; it is 0 for all other values.

4.5 Expectation, Moments, and Variance

Let us now look at finding the average or expected value of a random variable. Recall that earlier in this chapter a random variable is the mapping of the elements of the sample space into real numbers. Therefore, given a sequence of numbers say 2, 3, 4, 2, 1, 4, 5, 5, 1, 1, 2, 1, 3, 2... etc. Intuitively, to find the average, we sum up all of the numbers and divide by the total number of items as follows:

$$\text{average} = \frac{1n_1 + 2n_2 + 3n_3 + \ldots + kn_k}{N}$$ where n_1 is the number of 1s n_2 is the number of 2s etc. N is the total number items. Notice that the following grouping can be made:

$$\text{average} = 1\frac{n_1}{N} + 2\frac{n_2}{N} + 3\frac{n_3}{N} + \ldots + k\frac{n_k}{N}.$$ Notice that for large N n_k/N is the probability that the random variable equals k. Thus, we can write the average or expected value of X as:

$$[63] \quad \bar{X} = E[X] = \sum_i x_i p_i \text{ where } p_i = P[X = x_i], \text{ (discrete case)}$$

$\bar{X}$ is used to represent the expected value of X. E[X] is the expectation operator.

Example 4.10:

Refer back to example 4.5. What is the expected value of 1s that we could expect to see? Using [63] we get:

$$\bar{X} = E[X] = \sum_{i=0}^{3} x_i p_i = 0 \cdot \frac{1}{8} + 1 \cdot \frac{3}{8} + 2 \cdot \frac{3}{8} + 3 \cdot \frac{1}{8} = \frac{12}{8} = \frac{3}{2} \ \blacksquare$$

Adapting [63] for continuous random variables, we have the following as the expected value:

$$[64] \quad \bar{X} = E[X] = \int_{-\infty}^{\infty} x f_X(x) dx, \text{ (continuous case)}[1]$$

We can also find the expected value of X^n. This is called the nth moment of X and is defined below:

[65] $\overline{X^n} = E[X^n] = \sum_i x_i^n p_i$, (discrete case)

[66] $\overline{X^n} = E[X^n] = \int_{-\infty}^{\infty} x^n f_X(x)dx$, (continuous case)

The second moment of X is very important and is defined below. It represents the mean square value of X.

[67] $\overline{X^2} = E[X^2] = \sum_i x_i^2 p_i$, (discrete case)

[68] $\overline{X^2} = E[X^2] = \int_{-\infty}^{\infty} x^2 f_X(x)dx$, (continuous case)

A function of X, g(X), itself is a random variable, therefore, the expected value of a function of X can be found in a similar manner as in [64] and is defined as:

[69] $\overline{g(X)} = E[g(X)] = \int_{-\infty}^{\infty} g(x) f_X(x)dx$, (continuous case)

Another important quantity that is used when working with random variables is called the central moment and is defined as:

[70] $E[(X - \overline{X})^n] = \sum_i (x_i - \overline{X})^n p_i$, (discrete case)

[71] $E[(X - \overline{X})^n] = \int_{-\infty}^{\infty} (x - \overline{X})^n f_X(x)dx$, (continuous case)

---

1. This will also work if $f_X(x)$ is discrete.

In fact, the 2nd central moment is important enough to get its own name, variance. It is represented by the σ^2 symbol as shown below.

[72] $\sigma^2 = E[(X-\bar{X})^2]$

A useful relationship to calculate the variance can be found by expanding [72] as follows:

$$\sigma^2 = E[(X-\bar{X})^2] = E[X^2 - 2X\bar{X} + \bar{X}^2]$$

$= E[X^2] - E[2X\bar{X}] + E[\bar{X}^2] = \overline{X^2} - 2\bar{X}^2 + \bar{X}^2 = \overline{X^2} - \bar{X}^2$. Note that this can be done because the expected value of an expected value (i.e. a constant) is just the expected value. To summarize:

[73] $\sigma^2 = \overline{X^2} - \bar{X}^2$

Example 4.11:

Find the mean and variance of the following three cases:

a) c, c is a constant
b) X+c, c is a constant, X is a random variable
c) cX, c is a constant, X is a random variable

a) From the definition of expected value:

$$E[c] = \int_{-\infty}^{\infty} cf_X(x)dx = c\int_{-\infty}^{\infty} f_X(x)dx = c(1) = c$$

$$E[c^2] = \int_{-\infty}^{\infty} c^2 f_X(x)dx = c^2 \int_{-\infty}^{\infty} f_X(x)dx = c^2(1) = c^2$$

Using [73]:

variance $= E[c^2] - (E[c])^2 = c^2 - c^2 = 0$ as one might expect.

b)

$$E[X+c] = \int_{-\infty}^{\infty} (x+c)f_X(x)dx = \int_{-\infty}^{\infty} xf_X(x)dx + c\int_{-\infty}^{\infty} f_X(x)dx$$

$$= \overline{X} + c$$

$$E[(X+c)^2] = \int_{-\infty}^{\infty} (x+c)^2 f_X(x)dx = \int_{-\infty}^{\infty} (x^2 + 2cx + c^2)f_X(x)dx$$

$$= \int_{-\infty}^{\infty} x^2 f_X(x)dx + \int_{-\infty}^{\infty} 2cx f_X(x)dx + \int_{-\infty}^{\infty} c^2 f_X(x)dx = \overline{X^2} + 2c\overline{X} + c^2$$

Using [73]:

$$\text{variance} = E[(X+c)^2] - (E[X+c])^2 = \overline{X^2} + 2c\overline{X} + c^2 - (\overline{X}+c)^2$$

$\overline{X^2} + 2c\overline{X} + c^2 - \overline{X}^2 - 2c\overline{X} - c^2 = \overline{X^2} - \overline{X}^2 = \sigma^2$ which is just the variance of the random variable X.

c)

$$E[cX] = \int_{-\infty}^{\infty} cx f_X(x)dx = c\int_{-\infty}^{\infty} x f_X(x)dx = c\overline{X}$$

$$E[(cX)^2] = \int_{-\infty}^{\infty} (cx)^2 f_X(x)dx = c^2\int_{-\infty}^{\infty} x^2 f_X(x)dx = c^2\overline{X^2}$$

Using [73]:

$$\sigma^2 = E[(cX)^2] - (E[cX])^2 = c^2\overline{X^2} - (c\overline{X})^2 = c^2\overline{X^2} - c^2\overline{X}^2$$

$c^2(\overline{X^2} - \overline{X}^2) = c^2\sigma^2$ This is the variance times the square of the constant. ∎

The square root of the variance is called the standard deviation as shown below.

[74] Standard Deviation $= \sigma = \sqrt{\sigma^2}$

The standard deviation is sometimes used because it has the same units as its corresponding random variable.

Example 4.12:

Given the cdf function below, find the:

a) pdf
b) mean
c) mean square value
d) variance

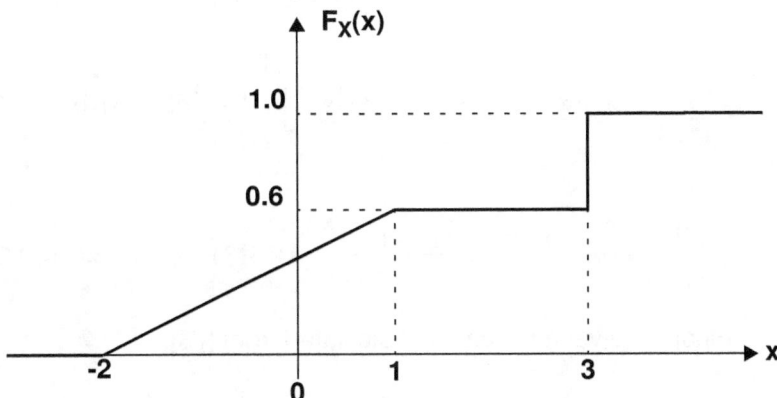

Figure 42 cdf for example 4.12

a) From [58] the pdf is simply the derivative of the cdf:

$$f_X(x) = \frac{d}{dx}F_X(x).$$

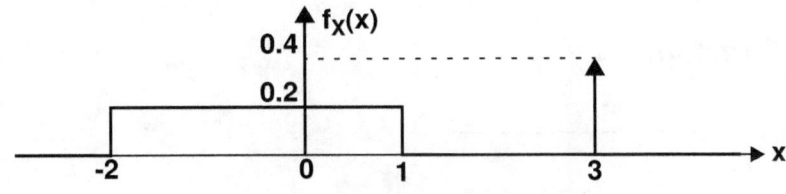

Figure 43 pdf of cdf in Figure 42

Notice that this pdf contains a mix of both a continuous portion and an impulse. A random variable which produces this type of cdf is called a mixed random variable. Also notice that the area under this pdf is 1 as is required for pdfs.

b) The mean can be found from [64]:

$$\overline{X} = \int_{-\infty}^{\infty} x f_X(x) dx = \int_{-2}^{1} 0.2x \, dx + \int_{2}^{4} 0.4x \delta(x-3) dx$$

$$= 0.2 \frac{x^2}{2}\bigg|_{-2}^{1} + 0.4x\big|_3 = 0.2 \frac{(1^2 - (-2)^2)}{2} + 0.4(3) = -0.3 + 1.2 = 0.9$$

c) The mean square value can be found from [68][1]:

$$\overline{X^2} = \int_{-\infty}^{\infty} x^2 f_X(x) dx = \int_{-2}^{1} 0.2x^2 \, dx + \int_{2}^{4} 0.4x^2 \delta(x-3) dx$$

$$= 0.2 \frac{x^3}{3}\bigg|_{-2}^{1} + 0.4x^2\big|_3 = 0.2 \frac{(1^3 - (-2)^3)}{3} + 0.4(3)^2 = 0.6 + 3.6 = 4.2$$

d) Finally the variance can be calculated from [73]:

$$\sigma^2 = \overline{X^2} - \overline{X}^2 = 4.2 - 0.9^2 = 3.39 \blacksquare$$

4.6 Density Functions of a Few Important Random Variables

In the sections below, we discuss some of the more important distribution functions. [1], [3], [4] and [9] from the bibliography list out a number of additional distributions and their properties.

4.6.1 Uniform

---

1. Note that $\int_{-\infty}^{\infty} g(x)\delta(x-a)dx = g(a)$.

The uniform density function has a constant, i.e. uniform, value as shown below.

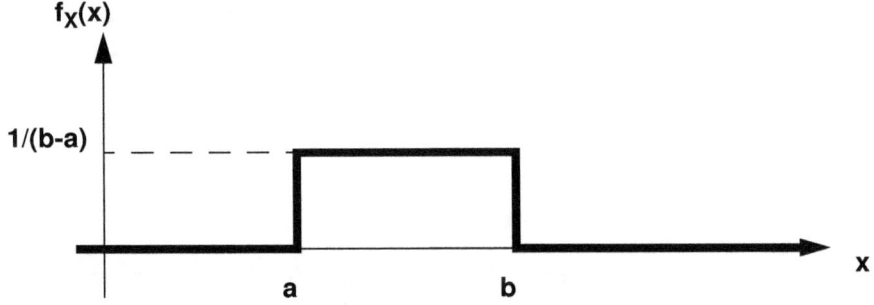

Figure 44 Uniform Density Function

The mean value is:

$$[75] \quad E[X] = \int_{-\infty}^{\infty} x f_X(x) dx = \frac{1}{b-a} \int_a^b x dx = \frac{1}{b-a} \left. \frac{x^2}{2} \right|_a^b = \frac{b^2 - a^2}{2(b-a)}$$

$$= \frac{(b-a)(b+a)}{2(b-a)} = \frac{b+a}{2}$$

$$[76] \quad E[X^2] = \int_{-\infty}^{\infty} x^2 f_X(x) dx = \frac{1}{b-a} \int_a^b x^2 dx = \frac{1}{b-a} \left. \frac{x^3}{3} \right|_a^b = \frac{b^3 - a^3}{3(b-a)}$$

$$= \frac{(b-a)(b^2 + ab + a^2)}{3(b-a)} = \frac{(b^2 + ab + a^2)}{3}$$

The variance is: $\sigma^2 = E[X^2] - (E[X])^2 = \frac{b^2 + ab + a^2}{3} - \left(\frac{b+a}{2}\right)^2$

$$= \frac{b^2 + ab + a^2}{3} - \frac{b^2 + a^2 + 2ab}{4} = \frac{4b^2 + 4ab + 4a^2 - 3b^2 - 3a^2 + -6ab}{12}$$

$$\frac{b^2 - 2ab + a^2}{12} = \frac{(b-a)^2}{12}$$

4.6.2 Gaussian

The gaussian random variable is very important due to the fact that it can be used to represent many physical situations. Its pdf is mathematically represented by the formula:

$$[77] \quad f_X(x) = \frac{e^{\dfrac{-(x-\overline{X})^2}{2\sigma^2}}}{\sqrt{2\pi}\sigma} \quad \text{for } -\infty < x < \infty$$

The figure below shows a plot of the cdf and pdf. Notice that the mean $\overline{X}$ occurs at the maximum of the pdf and that the pdf is an even function about the mean.

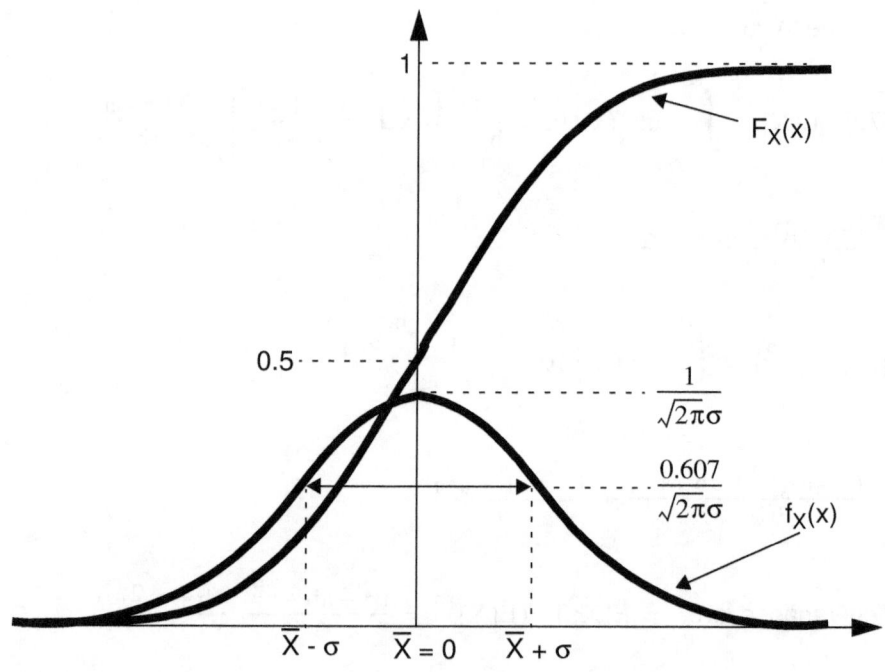

Figure 45 Plot of cdf and pdf of Gaussian Random Variable with Zero Mean and Variance of σ^2

The cdf is $F_X(x) = \int_{-\infty}^{x} \dfrac{e^{\frac{-(\xi - \overline{X})^2}{2\sigma^2}}}{\sqrt{2\pi}\sigma} d\xi$ and cannot be written in closed form.

However, it can be tabulated using numerical methods with $\overline{X}=0$ and $\sigma^2 =1$. We can then write the cdf as:

$$[78] \quad G(x) = \int_{-\infty}^{x} \dfrac{e^{\frac{-\xi^2}{2}}}{\sqrt{2\pi}} d\xi$$

Table 2 lists the tabulated values for several values of x. By using a change of variable where we set $\xi = (u-\overline{X})/\sigma$ and $d\xi=du/\sigma$ and x is set to $(x-\overline{X})/\sigma$ we get[1]:

$$[79] \quad G\left(\dfrac{x-\overline{X}}{\sigma}\right) = \int_{-\infty}^{\left(\frac{x-\overline{X}}{\sigma}\right)} \dfrac{e^{\frac{-(u-\overline{X})^2}{2\sigma^2}}}{\sqrt{2\pi}\sigma} du$$

Comparing [78] and [79] we can see that they are equivalent. Therefore, if we know $\overline{X}$ and σ then we can use $(x-\overline{X})/\sigma$ as the argument of G(x) and simply look up the value in Table 2.   If $(x-\overline{X})/\sigma$ turns out to be negative, the following relationship can be used:

[80] $G(-x) = 1 - G(x)$

Example 4.13:

A voltage with a gaussian distribution is applied to the input of a logic gate with a switching threshold of 0.45 volts. $\overline{X}= 0.3$ and $\sigma = 0.1$. What is the probability that the voltage will not exceed the switching threshold of the logic gate?

What is being asked for is $F_X(0.45\text{volts})=P(x \leq 0.45\text{volts})$. To find this we use:

$$G\left(\dfrac{x-\overline{X}}{\sigma}\right) = G\left(\dfrac{0.45-0.3}{0.1}\right) = G(1.5) = 0.9332$$

---

1. Remember that ξ is just a dummy variable for x, and in [79] u then becomes the dummy variable for x.

Additionally, the probability that the voltage will exceed the switching threshold of the logic gate is simply 1-0.9332 = 0.0668.∎

Table 2 G((x-X̄)/σ)

	0.00	0.01	0.02	0.03	0.04	0.05	0.06	0.07	0.08	0.09
0.0	0.5000	0.5040	0.5080	0.5120	0.5160	0.5199	0.5239	0.5279	0.5319	0.5359
0.1	0.5398	0.5438	0.5478	0.5517	0.5557	0.5596	0.5636	0.5675	0.5714	0.5753
0.2	0.5793	0.5832	0.5871	0.5910	0.5948	0.5987	0.6026	0.6064	0.6103	0.6141
0.3	0.6179	0.6217	0.6255	0.6293	0.6331	0.6368	0.6406	0.6443	0.6480	0.6517
0.4	0.6554	0.6591	0.6628	0.6664	0.6700	0.6736	0.6772	0.6808	0.6844	0.6879
0.5	0.6915	0.6950	0.6985	0.7019	0.7054	0.7088	0.7123	0.7157	0.7190	0.7224
0.6	0.7257	0.7291	0.7324	0.7357	0.7389	0.7422	0.7454	0.7486	0.7517	0.7549
0.7	0.7580	0.7611	0.7642	0.7673	0.7704	0.7734	0.7764	0.7794	0.7823	0.7852
0.8	0.7881	0.7910	0.7939	0.7967	0.7995	0.8023	0.8051	0.8078	0.8106	0.8133
0.9	0.8159	0.8186	0.8212	0.8238	0.8264	0.8289	0.8315	0.8340	0.8365	0.8389
1.0	0.8413	0.8438	0.8461	0.8485	0.8508	0.8531	0.8554	0.8577	0.8599	0.8621
1.1	0.8643	0.8665	0.8686	0.8708	0.8729	0.8749	0.8770	0.8790	0.8810	0.8830
1.2	0.8849	0.8869	0.8888	0.8907	0.8925	0.8944	0.8962	0.8980	0.8997	0.9015
1.3	0.9032	0.9049	0.9066	0.9082	0.9099	0.9115	0.9131	0.9147	0.9162	0.9177
1.4	0.9192	0.9207	0.9222	0.9236	0.9251	0.9265	0.9279	0.9292	0.9306	0.9319
1.5	0.9332	0.9345	0.9357	0.9370	0.9382	0.9394	0.9406	0.9418	0.9429	0.9441
1.6	0.9452	0.9463	0.9474	0.9484	0.9495	0.9505	0.9515	0.9525	0.9535	0.9545
1.7	0.9554	0.9564	0.9573	0.9582	0.9591	0.9599	0.9608	0.9616	0.9625	0.9633
1.8	0.9641	0.9649	0.9656	0.9664	0.9671	0.9678	0.9686	0.9693	0.9699	0.9706
1.9	0.9713	0.9719	0.9726	0.9732	0.9738	0.9744	0.9750	0.9756	0.9761	0.9767
2.0	0.9772	0.9778	0.9783	0.9788	0.9793	0.9798	0.9803	0.9808	0.9812	0.9817
2.1	0.9821	0.9826	0.9830	0.9834	0.9838	0.9842	0.9846	0.9850	0.9854	0.9857
2.2	0.9861	0.9864	0.9868	0.9871	0.9875	0.9878	0.9881	0.9884	0.9887	0.9890
2.3	0.9893	0.9896	0.9898	0.9901	0.9904	0.9906	0.9909	0.9911	0.9913	0.9916
2.4	0.9918	0.9920	0.9922	0.9925	0.9927	0.9929	0.9931	0.9932	0.9934	0.9936
2.5	0.9938	0.9940	0.9941	0.9943	0.9945	0.9946	0.9948	0.9949	0.9951	0.9952
2.6	0.9953	0.9955	0.9956	0.9957	0.9959	0.9960	0.9961	0.9962	0.9963	0.9964
2.7	0.9965	0.9966	0.9967	0.9968	0.9969	0.9970	0.9971	0.9972	0.9973	0.9974
2.8	0.9974	0.9975	0.9976	0.9977	0.9977	0.9978	0.9979	0.9979	0.9980	0.9981
2.9	0.9981	0.9982	0.9982	0.9983	0.9984	0.9984	0.9985	0.9985	0.9986	0.9986
3.0	0.9987	0.9987	0.9987	0.9988	0.9988	0.9989	0.9989	0.9989	0.9990	0.9990
3.1	0.9990	0.9991	0.9991	0.9991	0.9992	0.9992	0.9992	0.9992	0.9993	0.9993
3.2	0.9993	0.9993	0.9994	0.9994	0.9994	0.9994	0.9994	0.9995	0.9995	0.9995
3.3	0.9995	0.9995	0.9995	0.9996	0.9996	0.9996	0.9996	0.9996	0.9996	0.9997
3.4	0.9997	0.9997	0.9997	0.9997	0.9997	0.9997	0.9997	0.9997	0.9997	0.9998
3.5	0.9998	0.9998	0.9998	0.9998	0.9998	0.9998	0.9998	0.9998	0.9998	0.9998
3.6	0.9998	0.9998	0.9999	0.9999	0.9999	0.9999	0.9999	0.9999	0.9999	0.9999
3.7	0.9999	0.9999	0.9999	0.9999	0.9999	0.9999	0.9999	0.9999	0.9999	0.9999
3.8	0.9999	0.9999	0.9999	0.9999	0.9999	0.9999	0.9999	0.9999	0.9999	0.9999

Example 4.14:

A voltage with a gaussian distribution is applied to the circuit below:

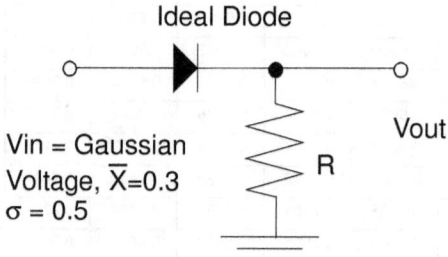

Ideal Diode

Vin = Gaussian
Voltage, $\overline{X}$=0.3
$\sigma = 0.5$

R

Vout

Figure 46 Circuit for example 4.14

What will the pdf of Vout look like?

Any voltage under 0 volts at Vin will appear as 0 volts at Vout, other-wise Vout = Vin. Therefore, the pdf of Vout will be 0 for voltages less than 0. However, at 0 volts there will be an impulse in the pdf of Vout to represent the probability that Vout will be 0. This probability is the probability that Vin ≤ 0volts and can be found using G((0 - 0.3) / 0.5) = G(-0.60). Using [80] and Table 2: G(-x) = 1 - G(x) = 1 - G(0.60) = 1 - 0.7257 = 0.2743, which is the value of the impulse. Since Vout = Vin when Vin > 0volts the rest of the pdf is unchanged. The figure below shows the pdf of Vout.

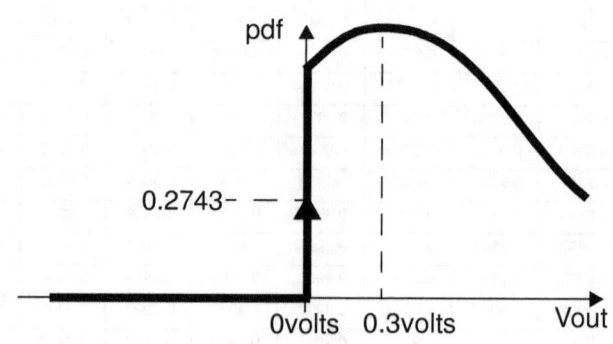

pdf

0.2743- —

0volts   0.3volts   Vout

Figure 47 pdf of Vout

■

70

Next we shall derive an expression for the nth moment of a zero mean gaussian random variable. We start off with the relation:

$$\int_0^\infty e^{-a^2 x^2} dx = \frac{\sqrt{\pi}}{2a}.$$

Due to the pdf's symmetry, we can extend the range of integration as follows:

[81] $\int_{-\infty}^\infty e^{-a^2 x^2} dx = 2\int_0^\infty e^{-a^2 x^2} dx = \frac{\sqrt{\pi}}{a}$

Therefore:

[82] $\int_{-\infty}^\infty e^{-a^2 x^2} dx = \sqrt{\frac{\pi}{a}}$

Using Leibnitz's Rule[1] we differentiate both sides of [82] repeatedly as shown below (three differentiations are shown, but more can be done):

$$\frac{d}{da}\int_{-\infty}^\infty e^{-a^2 x^2} dx = -\int_{-\infty}^\infty x^2 e^{-a^2 x^2} dx = \frac{d}{da}\sqrt{\frac{\pi}{a}} = -\frac{\sqrt{\pi}}{2a^{3/2}}$$

$$-\frac{d}{da}\int_{-\infty}^\infty x^2 e^{-a^2 x^2} dx = \int_{-\infty}^\infty x^4 e^{-a^2 x^2} dx = -\frac{d}{da}\frac{\sqrt{\pi}}{2a^{3/2}} = \frac{3\sqrt{\pi}}{4a^{5/2}}$$

$$\frac{d}{da}\int_{-\infty}^\infty x^4 e^{-a^2 x^2} dx = -\int_{-\infty}^\infty x^6 e^{-a^2 x^2} dx = \frac{d}{da}\frac{3\sqrt{\pi}}{4a^{5/2}} = \frac{-15\sqrt{\pi}}{8a^{7/2}}$$

etc.,

If we now substitute $1/(2\sigma^2)$ for a in the above equations we get:

---

1. $\frac{\partial}{\partial\lambda}\int_a^b f(x, \lambda)dx = \int_a^b \frac{\partial}{\partial\lambda}f(x, \lambda)dx$.

$$\int_{-\infty}^{\infty} x^2 e^{-\frac{x^2}{2\sigma^2}} dx = \frac{\sqrt{\pi}}{2\left(\frac{1}{2\sigma^2}\right)^{3/2}} = \sqrt{2\pi}\sigma^3$$

$$\int_{-\infty}^{\infty} x^4 e^{-\frac{x^2}{2\sigma^2}} dx = \frac{3\sqrt{\pi}}{4\left(\frac{1}{2\sigma^2}\right)^{5/2}} = 3\sqrt{2\pi}\sigma^5$$

$$\int_{-\infty}^{\infty} x^6 e^{-\frac{x^2}{2\sigma^2}} dx = \frac{15\sqrt{\pi}}{8\left(\frac{1}{2\sigma^2}\right)^{7/2}} = 15\sqrt{2\pi}\sigma^7$$

Then dividing both sides of the equations by $(2\pi)^{1/2}\sigma$, we recognize the nth moment of the gaussian random variable as shown below:

$$\int_{-\infty}^{\infty} \frac{x^2 e^{-\frac{x^2}{2\sigma^2}}}{\sqrt{2\pi}\sigma} dx = \sigma^2 \quad \text{The 2nd moment is just } 1*\sigma^2.$$

$$\int_{-\infty}^{\infty} \frac{x^4 e^{-\frac{x^2}{2\sigma^2}}}{\sqrt{2\pi}\sigma} dx = 3\sigma^4 \quad \text{The 4th moment is just } 1*3*\sigma^4.$$

$$\int_{-\infty}^{\infty} \frac{x^6 e^{-\frac{x^2}{2\sigma^2}}}{\sqrt{2\pi}\sigma} dx = 15\sigma^6 \quad \text{The 6th moment is just } 1*3*5*\sigma^6.$$

Thus, a pattern is recognized; the nth moment is $1*3*5*...(n-1)*\sigma^n$ for when n is even. Let us next consider the case when n is odd. First note that the pdf of a zero mean gaussian random variable is an even function[1] about x=0. Note that, as a result, we can write:

---

1. An even function is one in which f(x)=f(-x).

[83] $\displaystyle \int_{-\infty}^{\infty} x^n f_X(x)dx = \int_{-\infty}^{0} x^n f_X(x)dx + \int_{0}^{\infty} x^n f_X(x)dx$

Since the pdf is an even function, if we multiply it by x^n where n is odd, we turn $x^n f_X(x)$ into an odd function[1]. This allows us to write:

[84] $\displaystyle \int_{-\infty}^{0} x^n f_X(x)dx = -\int_{0}^{\infty} x^n f_X(x)dx$

Thus inserting [84] into [83] we get: $\displaystyle -\int_{0}^{\infty} x^n f_X(x)dx + \int_{0}^{\infty} x^n f_X(x)dx = 0$,

for n odd.

The results are summarized below:

[85] $\overline{X^n} = 0$ for n odd

[86] $\overline{X^n} = 1 \bullet 3 \bullet 5 \bullet \ldots(n-1)\sigma^n$ for n even

Before leaving our discussion of gaussian random variables, we mention two additional tools used for calculating the cdf to make you aware of their existence. These are called the Q function and error function. The Q function is defined as:

[87] $Q\left(\dfrac{x-\overline{X}}{\sigma}\right) = 1 - G\left(\dfrac{x-\overline{X}}{\sigma}\right)$

The error function[2] is defined in terms of the Q function as:

[88] $Q(x) = \dfrac{1}{2}\left(1 - \text{erf}\left(\dfrac{x}{\sqrt{2}}\right)\right)$

An approximation of the Q function is given in reference [4] page 70, which provides a good approximation when a $\geq$ 1.25.

---

1. An odd function is defined as: $f(-x) \equiv -f(x)$.
2. erf(x) is used to represent the error function.

4.6.3 Exponential

The cdf and pdf for the exponential random variable are given below:

[89] $\quad F_T(\tau) = 1 - e^{-\frac{\tau}{\bar{\tau}}} \quad \tau \geq 0$

[90] $\quad f_T(\tau) = \frac{1}{\bar{\tau}} e^{-\frac{\tau}{\bar{\tau}}} \quad \tau \geq 0$

$\bar{\tau}$ is the average time of occurrence between events.

The exponential random variable is useful for making reliability calculations. For example, it can be used to determine the probability that a component will fail a certain amount of time before or after it's average expected lifetime.

The expected value is[1]:

$$E[\tau] = \int_0^\infty \xi \frac{1}{\bar{\tau}} e^{-\frac{\xi}{\bar{\tau}}} d\xi = -\xi e^{-\frac{\xi}{\bar{\tau}}} \Big|_0^\infty + \int_0^\infty e^{-\frac{\xi}{\bar{\tau}}} d\xi = -\xi e^{-\frac{\xi}{\bar{\tau}}} \Big|_0^\infty - \bar{\tau} e^{-\frac{\xi}{\bar{\tau}}} \Big|_0^\infty$$

$$= \left(-\xi e^{-\frac{\xi}{\bar{\tau}}} \Big|_\infty \right) - \left(-\xi e^{-\frac{\xi}{\bar{\tau}}} \Big|_0 \right) - \left(\bar{\tau} e^{-\frac{\xi}{\bar{\tau}}} \Big|_\infty \right) + \left(\bar{\tau} e^{-\frac{\xi}{\bar{\tau}}} \Big|_0 \right) = \left(-\xi e^{-\frac{\xi}{\bar{\tau}}} \Big|_\infty \right) - 0 - 0 + \bar{\tau} = \bar{\tau}$$

Note that $\left(-\xi e^{-\frac{\xi}{\bar{\tau}}} \Big|_\infty \right)$ reduces to 0 because $e^{-\frac{\xi}{\bar{\tau}}}$ approaches 0 faster than ξ approaches infinity.

---

1. Integration by parts is used: $\int u dv = uv - \int v du$ where $u = \xi$, thus $du = d\xi$, and $dv = (1/\bar{\tau})e^{-\xi/\bar{\tau}}d\xi$ thus $v = -e^{-\xi/\bar{\tau}}$.

Next we calculate the mean square value where, again, integration by parts is used[1]. $u = \xi^2$ thus $du = 2\xi d\xi$, and $dv = (1/\tau)e^{-\xi/\tau}$. Thus $v = -e^{-\xi/\tau}d\xi$.

$$E[\tau^2] = \int_0^\infty \xi^2 \frac{1}{\tau} e^{-\frac{\xi}{\tau}} d\xi = -\xi^2 e^{-\frac{\xi}{\tau}}\Big|_0^\infty + \int_0^\infty 2\xi e^{-\frac{\xi}{\tau}} d\xi$$

$$= -\xi^2 e^{-\frac{\xi}{\tau}}\Big|_0^\infty - 2\bar{\tau}^2 e^{-\frac{\xi}{\tau}}\Big|_0^\infty - 2\xi e^{-\frac{\xi}{\tau}}\Big|_0^\infty = \left(-\xi^2 e^{-\frac{\xi}{\tau}}\Big|_\infty\right) - \left(-\xi^2 e^{-\frac{\xi}{\tau}}\Big|_0\right)$$

$$-\left(2\bar{\tau}^2 e^{-\frac{\xi}{\tau}}\Big|_\infty\right) + \left(2\bar{\tau}^2 e^{-\frac{\xi}{\tau}}\Big|_0\right) - \left(2\xi e^{-\frac{\xi}{\tau}}\Big|_\infty\right) + \left(2\xi e^{-\frac{\xi}{\tau}}\Big|_0\right)$$

$$= 0 - 0 - 0 + 2\bar{\tau}^2 - 0 + 0 = 2\bar{\tau}^2$$

Note that $\left(-\xi^2 e^{-\frac{\xi}{\tau}}\Big|_\infty\right)$ and $\left(2\xi e^{-\frac{\xi}{\tau}}\Big|_\infty\right)$ both reduce to 0 because $e^{-\frac{\xi}{\tau}}$ approaches 0 faster than ξ^2 or 2ξ approach infinity[2].

The variance can now be calculated from [73]:

$$\sigma^2 = E[\tau^2] - (E[\tau])^2 = 2\bar{\tau}^2 - (\bar{\tau})^2 = \bar{\tau}^2$$

Example 4.15:

A component has an operating life that is exponentially distributed with a mean time to failure of 5 years. What is:

---

1. The relationship $\int xe^{\frac{-x}{b}} dx = -\left(b^2 e^{\frac{-x}{b}} + xe^{\frac{-x}{b}}\right)$ is used to help carry out the integrations.
2. L'Hôpital's rule could have also been used to evaluate these indeterminate forms. See reference [10], page 119, equation 4.7-9.

a) The probability that the component will fail in the first year of operation,

b) The probability that the component will fail after 15 years of operation, and

c) The probability that the component will fail between 3 and 7 years during operation.

a) The probability of failing within the first year is $F_X(1)=P(X \leq 1) = 1 - e^{-1/5} = 0.181$.

b) To find the probability that the component will fail after 15 years, we first find the probability that the component will fail before 15 years and subtract that value from 1; see equation [57]. Therefore, $P(X>15) = 1 - P(X \leq 15) = 1 - (1 - e^{-15/5}) = 0.05$.

c) Using [56], we get $P(3 < X \leq 7) = F_X(7) - F_X(3) = 1 - e^{-7/5} - 1 + e^{-3/5} = 0.302$. ∎

4.7   Conditional cdf, pdf, and Expectation

Conditional probability was defined by [30] in section 2.5. This definition can be extended to cdfs as follows:

$$[91] \quad F_X(X \leq x | X \leq m) = \frac{P(X \leq x \cap X \leq m)}{P(X \leq m)}$$

The properties of conditional cdfs are similar to those for unconditional cdfs[1].

$[92] \quad 0 \leq F_X(X \leq x \mid X \leq m) \leq 1 \quad -\infty < x < \infty$

$[93] \quad F_X(X = -\infty \mid X \leq m) = 0$

$[94] \quad F_X(X = \infty \mid X \leq m) = 1$

$[95] \quad F_X(X \leq x \mid X \leq m)$ is non-decreasing as x increases

$[96] \quad F_X(x_1 < X \leq x_2 \mid X \leq m) = F_X(X \leq x_2 \mid X \leq m) - F_X(X \leq x_1 \mid X \leq m)$ for $x_2 > x_1$.

Furthermore, if $x > m$ then $P(X \leq x \cap X \leq m) = P(X \leq m)$. Therefore, [91] becomes:

$$[97] \quad F_X(X \leq x | X \leq m) = \frac{P(X \leq x \cap X \leq m)}{P(X \leq m)} = \frac{P(X \leq m)}{P(X \leq m)} = 1, \quad \text{for } x \leq m$$

---

1. The properties shown here are for conditional cdfs that are conditioned on $x \leq m$. The properties for conditional cdfs that are conditioned on $x > m$ can be obtained by replacing $x \leq m$ with $x > m$.

If x ≤ m then [91] becomes:

$$[98] \quad F_X(X \le x | X \le m) = \frac{P(X \le x \cap X \le m)}{P(X \le m)} = \frac{P(X \le x)}{P(X \le m)} = \frac{F_X(x)}{F_X(m)}, \text{ for } x \le m$$

What [97] and [98] are showing is that for a conditional cdf, once x exceeds m, the probability that x > m is 1 and when x ≤ m then the cdf is just a scaled version of the unconditional cdf as shown in the figure below.

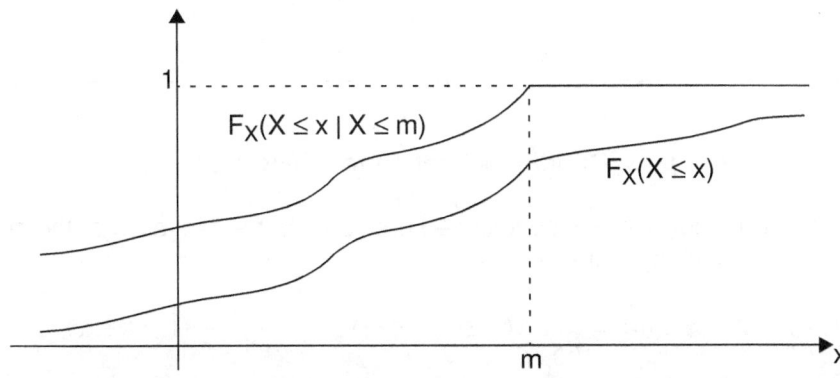

Figure 48 Conditional cdf conditioned on x ≤ m

If we were instead to condition on x > m [91] becomes:

$$[99] \quad F_X(X \le x | X > m) = \frac{P(X \le x \cap X > m)}{P(X > m)} = \frac{\varnothing}{P(X > m)} = 0, \text{ for } x \le m$$

If x > m then P(X ≤ x ∩ X > m) is the probability that X is between x and m; this is found by using [56]. Therefore, for x > m we can rewrite [91] as:

$$[100] \quad F_X(X \le x | X > m) = \frac{P(X \le x \cap X > m)}{P(X > m)} = \frac{P(X \le x) - P(X \le m)}{P(X > m)}$$

$$= \frac{F_X(x) - F_X(m)}{1 - F_X(m)}, \text{ for } x > m$$

The results of [99] and [100] are represented graphically by the figure below.

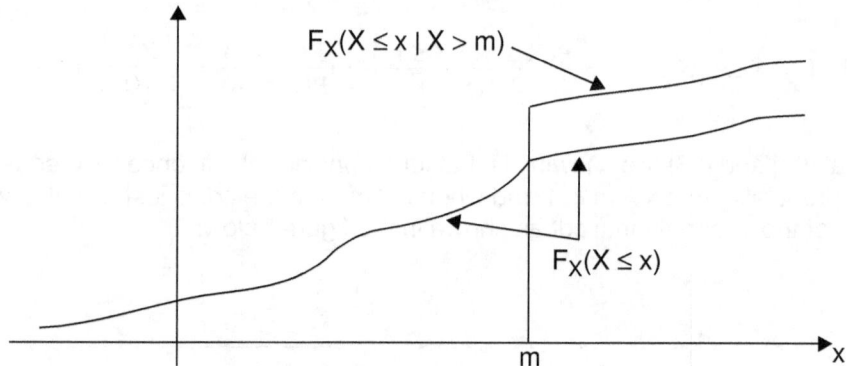

Figure 49 Conditional cdf conditioned on x > m

The conditional pdf which is conditioned on $x \leq m$ is the derivative of the corresponding conditional cdf as follows:

[101] $f_X(X \leq x | X \leq m) = \dfrac{d}{dx} F_X(X \leq x | X \leq m) = \dfrac{1}{F_X(X \leq m)} f_X(X \leq x)$

$$= \dfrac{f_X(X \leq x)}{\displaystyle\int_{-\infty}^{m} f_X(X \leq x) dx}, \text{ for } x < m$$

and

[102] $f_X(X \leq x | X \leq m) = 0, \text{ for } x \geq m$

Thus, the conditional pdf is a scaled version of the unconditional pdf for $x < m$ and is 0 for $x \geq m$ as shown in the figure below.

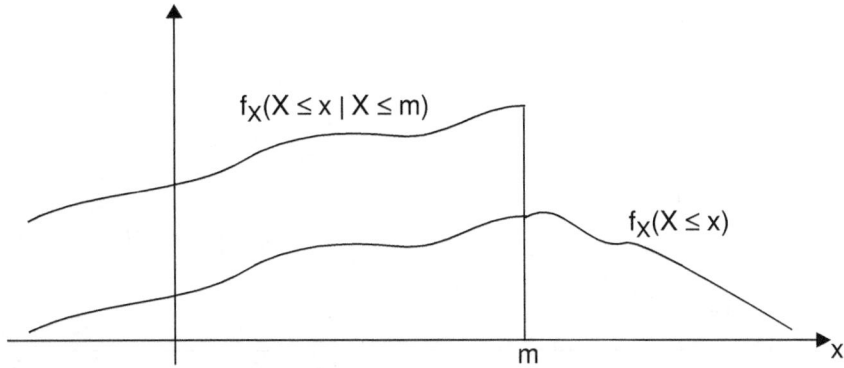

Figure 50 Conditional pdf conditioned on x ≤ m

If we are conditioning on $x > m$ then

[103] $f_X(X \le x | X > m) = \dfrac{d}{dx} F_X(X \le x | X > m) = \dfrac{1}{1 - F_X(m)} f_X(X \le x)$

$$= \frac{f_X(X \le x)}{1 - \displaystyle\int_{-\infty}^{m} f_X(X \le x) dx}, \text{ for } x > m$$

and

[104] $f_X(X \le x | X > m) = 0, \text{ for } x \le m$

The figure below graphically represents the results of [103] and [104].

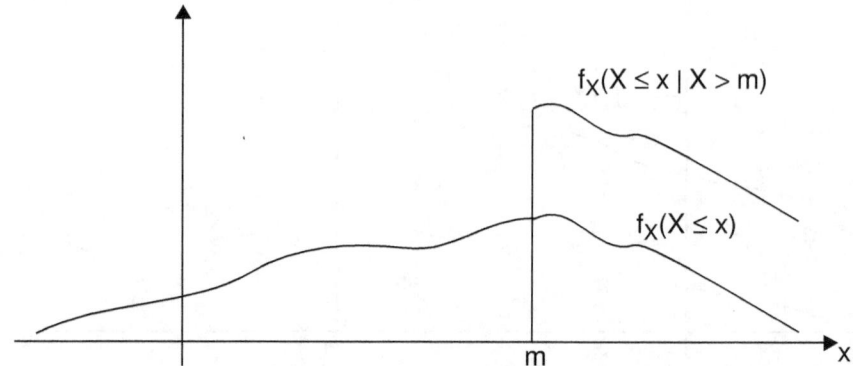

Figure 51 Conditional pdf conditioned on x > m

The properties of conditional pdfs are much like those for unconditional pdfs[1]:

[105] $f_X(X \le x \mid X \le m) \ge 0 \ \text{-}\ \infty < x < \infty$

[106] $\displaystyle\int_{-\infty}^{\infty} f_X(X \le x \mid X \le m)\,dx\ =\ 1$

[107] $F_X(X \le x \mid X \le m)\ =\ \displaystyle\int_{-\infty}^{x} f_X(X \le \xi \mid X \le m)\,d\xi$

[108] $\displaystyle\int_{x_1}^{x_2} f_X(X \le x \mid X \le m)\,dx\ =\ P(x_1 < X \le x_2 \mid X \le m)$

Conditional means also are similar to unconditional means. [69] can be adapted to use conditional pdfs as follows:

[109] $E[g(X) \mid X \le m]\ =\ \displaystyle\int_{-\infty}^{\infty} g(x)f_X(X \le x \mid X \le m)\,dx$

---

1. The properties shown here are for conditional pdfs that are conditioned on $x \le m$. The properties for conditional pdfs that are conditioned on $x > m$ can be obtained by replacing $x \le m$ with $x > m$. Also in [107] the range of integration is changed to m through ∞.

Example 4.16:

Let's start with a simple example. A voltage that is uniformly distributed from a to b is applied to a resistance of R ohms. What is the expected current if the voltage is above m volts.

For x > m the conditional pdf is found from [103]:

$$f_X(X \le x | X > m) = \frac{f_X(X \le x)}{1 - \int_{-\infty}^{m} f_X(X \le x)dx} = \frac{\frac{1}{b-a}}{1 - \frac{m-a}{b-a}} = \frac{\frac{1}{b-a}}{\frac{b-a-m+a}{b-a}}$$

$$= \frac{\frac{1}{b-a}}{\frac{b-m}{b-a}} = \frac{1}{b-m}$$

The figure below shows the conditional and unconditional pdfs.

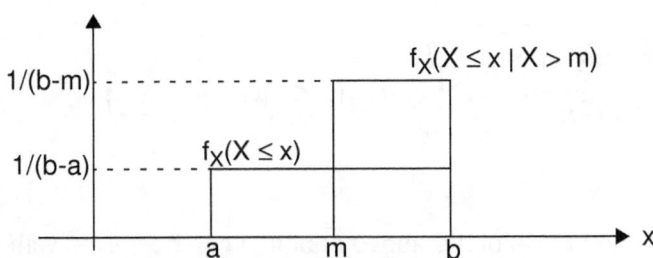

Figure 52 pdfs for example 4.16

The expected value of the current is then found by using [109] and swapping X ≤ m with X > m:

$$E[g(X)|X > m] = \int_{-\infty}^{\infty} g(x)f_X(X \le x | X > m)dx = \frac{1}{b-m}\int_{m}^{b} \frac{\xi}{R}d\xi$$

$$\frac{\xi^2}{2R(b-m)}\bigg|_m^b = \frac{b^2-m^2}{2R(b-m)} = \frac{(b-m)(b+m)}{2R(b-m)} = \frac{(b+m)}{2R} \quad \text{as intuitively}$$

would be expected from Figure 52. ∎

Example 4.17:

A voltage that has a gaussian distribution with 0 mean and standard deviation σ is passed through a half wave rectifier so that only voltages less than 0 are passed through.

a) What is the expected value of the voltage?
b) Can the results from part a be generalized if the input voltage has a non zero mean of $\overline{X}$?

a) The expected value of the voltage is found by using [109]:

$$E[X|X \le 0] = \int_{-\infty}^{0} x f_X(X \le x|X \le 0) dx = \int_{-\infty}^{0} \frac{x f_X(X \le x)}{F_X(0)} dx$$

$F_X(0)$ can be found by looking it up in Table 2 as 0.5. Therefore:

$$\int_{-\infty}^{0} \frac{x f_X(X \le x)}{F_X(0)} dx = 2\int_{-\infty}^{0} x f_X(X \le x) dx = \frac{2}{\sqrt{2\pi}\sigma} \int_{-\infty}^{0} x e^{-\frac{x^2}{2\sigma^2}} dx$$

The right hand side of the above equation can be solved with the aid of the following integral:

$$\int_{0}^{\infty} x e^{-ax^2} dx = \frac{1}{2a}$$

However, since the gaussian pdf is symmetrical about its mean, we can write this integral as:

$$\int_{0}^{\infty} xe^{-ax^2}dx = -\int_{-\infty}^{0} xe^{-ax^2}dx \Rightarrow \int_{-\infty}^{0} xe^{-ax^2}dx = -\frac{1}{2a} \text{ Thus,}$$

$$\frac{2}{\sqrt{2\pi}\sigma}\int_{-\infty}^{0} xe^{-\frac{x^2}{2\sigma^2}}dx = \frac{2}{\sqrt{2\pi}\sigma}\left(-\frac{1}{2\left(\frac{1}{2\sigma^2}\right)}\right) = -\frac{2\sigma^2}{\sqrt{2\pi}\sigma} = -\sqrt{\frac{2}{\pi}}\sigma$$

b)   The result above can be generalized for the case when the unconditional mean is not 0. Note that the conditional mean is always a fixed distance of $-(2/\pi)^{1/2}\sigma$ from the unconditional mean. Thus, if the unconditional mean shifts from 0 to $\overline{X}$ then the conditional mean also shifts by $\overline{X}$. Therefore, a general expression of the conditional mean is:

[110] $\overline{X} - \sqrt{\frac{2}{\pi}}\sigma$

One word of caution on the application of [110]. Its derivation assumed that the conditioning value was equal to the mean of the unconditional pdf. Therefore, [110] can only be used in situations where the conditional Gaussian pdf is cut in half about the mean of the unconditional Gaussian pdf. The figure below illustrates this point.

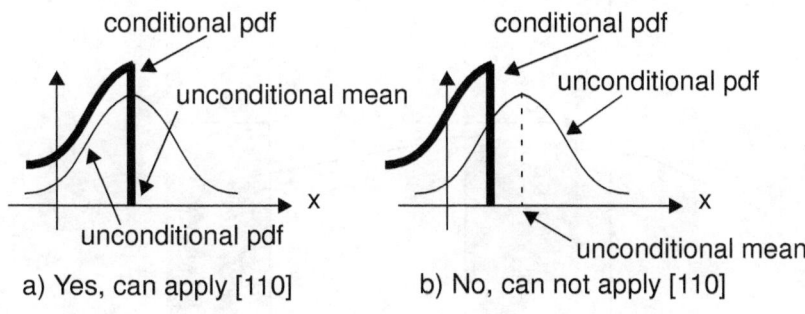

a) Yes, can apply [110]   b) No, can not apply [110]

Figure 53 When [110] can and can't be used

4.8 Determination of $f_Y(y)$ given $f_X(x)$ and $g(x)$

There may arise situations where a random variable X, with pdf, $f_X(x)$, is transformed into another random variable, Y, through a function, $g(x)$. In such a case it is desired to know the pdf of the new random variable Y. Here we show how to determine $f_Y(y)$. Consider Figure 54 below.

For this case there are two solutions to $y=g(x)$, namely x_1 and x_2. Using [61] we can write:

$$[111] \quad \int_{y_1}^{y_2} f_Y(\xi)d\xi = P(y_1 \leq Y \leq y_2)$$

Writing this differentially:

$$[112] \quad f_Y(y)dy = P(y \leq Y \leq y + dy)$$

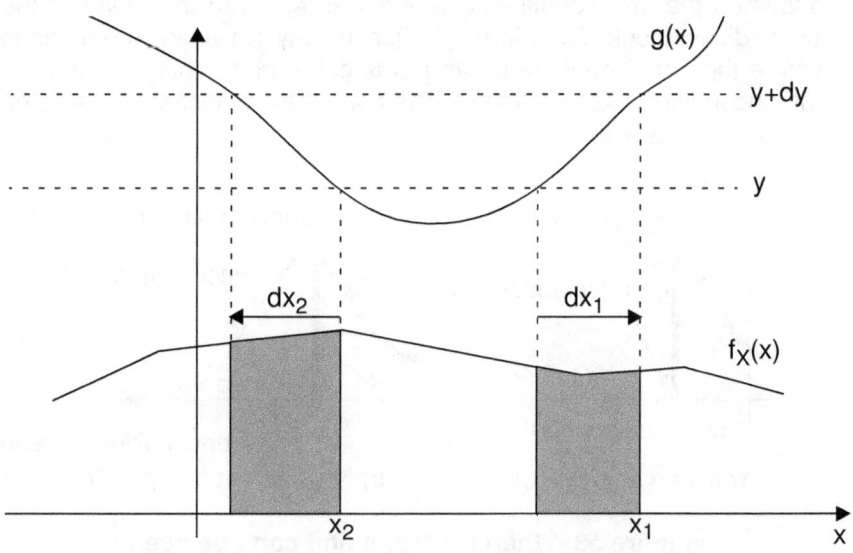

Figure 54 g(x) and Corresponding $f_X(x)$

Because the probability in [112] must equal the probability of the corresponding sets on the x-axis we can write:

[113] $P(y \leq Y \leq y + dy) = P(x_1 \leq X \leq x_1 + dx_1) + P(x_2 + dx_2 \leq X \leq x_2)$

Writing [113] differentially:

[114] $f_Y(y)dy = f_X(x_1)dx_1 + f_X(x_2)dx_2$

[115] Note that: $\dfrac{d}{dx}g(x_n) = \dfrac{dy}{dx_n} \Rightarrow dx_n = \dfrac{dy}{\dfrac{d}{dx}g(x_n)}$

Substituting [115] into [114]:

[116] $f_Y(y)dy = f_X(x_1)\dfrac{dy}{\dfrac{d}{dx}g(x_1)} + f_X(x_2)\dfrac{dy}{\left|\dfrac{d}{dx}g(x_2)\right|}$

The absolute value is taken of the derivative of $g(x_2)$ because the probability should be positive. The dy terms cancel out to give:

[117] $f_Y(y) = \dfrac{f_X(x_1)}{\dfrac{d}{dx}g(x_1)} + \dfrac{f_X(x_2)}{\left|\dfrac{d}{dx}g(x_2)\right|}$

This result can be generalized to:

[118] $f_Y(y) = \displaystyle\sum_{n=1}^{j} \dfrac{f_X(x_n)}{\left|\dfrac{d}{dx}g(x_n)\right|}$ Where j is the number of real roots of g(x)

If there are no real roots of g(x) then $f_Y(y)$ is 0 for the corresponding range of y.

There is a special case where g(x) is constant in a range from x_0 to x_1 as shown below.

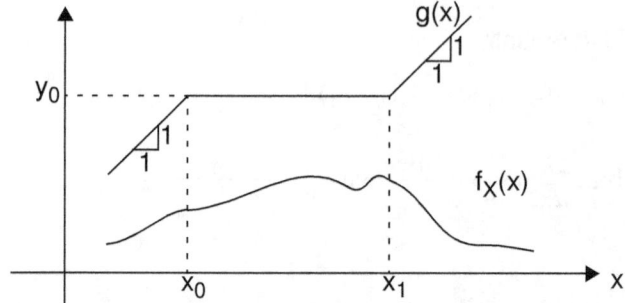

Figure 55 Case for g(x) is a constant

In this case [118] still applies, however, there will be an impulse function at $y=y_0$ that will be equal to $F_X(x_1) - F_X(x_0)$[1]. The resulting pdf for Figure 55 is shown below.

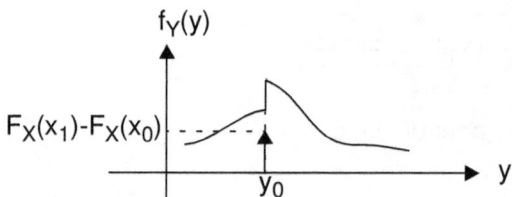

Figure 56 $f_Y(y)$ corresponding to Figure 55

Below are several examples which show how to apply the results above.

Example 4.18:

Given the system below, find $f_Y(y)$ and $\overline{Y}$.

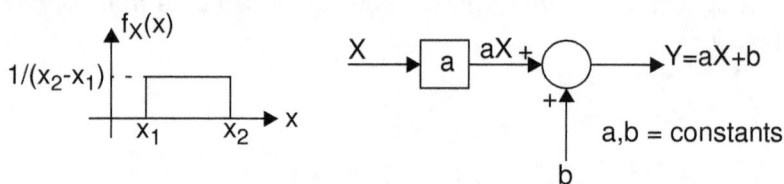

Figure 57 Example 4.18

---

1. Recall that $F_X(x_1) - F_X(x_0)$ is the area under $f_X(x)$ from x_0 to x_1.

There is only one root: x1=(y-b)/a. Also |dy/dx| = |a|. Therefore, using [118]:

$$f_Y(y) = \sum_{n=1}^{1} \frac{f_X(x_n)}{\left|\dfrac{d}{dx}g(x_n)\right|} = \frac{f_X\left(\dfrac{y-b}{a}\right)}{|a|} = \frac{1}{|a|(x_2 - x_1)}$$

Because $f_X(x)$ is a constant, inserting (y-b)/a into $f_X(x)$ does not change its value. The limits x_1 and x_2 transform into $y_1=ax_1+b$ and $y_2=ax_2+b$. Thus, the resulting pdf looks like:

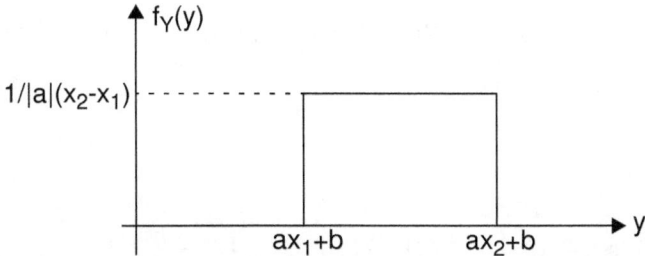

Figure 58 $f_Y(y)$ for example 4.18

Now we can find $\overline{Y}$ by:

$$\overline{Y} = \int_{-\infty}^{\infty} y f_Y(y)dy = \int_{ax_1+b}^{ax_2+b} \frac{y}{a(x_2-x_1)}dy = \left.\frac{y^2}{2a(x_2-x_1)}\right|_{ax_1+b}^{ax_2+b}$$

$$= \frac{a^2 x_2^2 + 2abx_2 + b^2 - (a^2 x_1^2 + 2abx_1 + b^2)}{2a(x_2-x_1)} = \frac{a^2(x_2^2 - x_1^2) + 2ab(x_2 - x_1)}{2a(x_2-x_1)}$$

$$= \frac{a(x_2-x_1)(x_2+x_1) + 2b(x_2-x_1)}{2(x_2-x_1)} = \frac{a(x_2+x_1) + 2b}{2} = \frac{a(x_2+x_1)}{2} + b$$

■

Example 4.19:

Find $f_Y(y)$ given Y=1/X.

There is one root $x_1 = 1/y_1$. $|dy/dx| = |-1/x^2| = 1/x^2$. Therefore, applying [118]:

$$f_Y(y) = \sum_{n=1}^{1} \frac{f_X(x_n)}{\left|\frac{d}{dx}g(x_n)\right|} = \left.\frac{f_X\left(\frac{1}{y}\right)}{\frac{1}{x_1^2}}\right|_{x_1 = \frac{1}{y^2}} = \frac{1}{y^2}f_X\left(\frac{1}{y}\right) \quad \blacksquare$$

Example 4.20:

Find $f_Y(y)$ for $Y=aX^2$. $|dy/dx| = |2ax|$. There are two real roots for $y \geq 0$, $\pm\sqrt{\frac{y}{a}}$. Applying [118]:

$$f_Y(y) = \sum_{n=1}^{2} \frac{f_X(x_n)}{\left|\frac{d}{dx}g(x_n)\right|} = \left.\frac{f_X\left(\sqrt{\frac{y}{a}}\right)}{|2ax_1|}\right|_{x_1 = \sqrt{\frac{y}{a}}} + \left.\frac{f_X\left(-\sqrt{\frac{y}{a}}\right)}{|2ax_2|}\right|_{x_2 = -\sqrt{\frac{y}{a}}}$$

$$= \frac{f_X\left(\sqrt{\frac{y}{a}}\right) + f_X\left(-\sqrt{\frac{y}{a}}\right)}{2\sqrt{ay}} \quad \blacksquare$$

Example 4.21:

Find $f_Y(y)$ for $Y=X^3$. $|dy/dx| = |3x^2|$. There is one solution $\sqrt[3]{y}$. Applying [118]:

$$f_Y(y) = \sum_{n=1}^{1} \frac{f_X(x_n)}{\left|\frac{d}{dx}g(x_n)\right|} = \left.\frac{f_X(\sqrt[3]{y})}{|3x_1^2|}\right|_{x_1 = \sqrt[3]{y}} = \frac{f_X(\sqrt[3]{y})}{3y^{2/3}} \quad \blacksquare$$

Example 4.22:

Find $f_Y(y)$ given $Y=X^{1/2}$. $|dy/dx| = 1/(2x^{1/2})$. There are no real roots for $y<0$. For $y\geq0$ there is one root: $x_1=y^2$. Applying [118]:

$$f_Y(y) = \sum_{n=1}^{1} \frac{f_X(x_n)}{\left|\frac{d}{dx}g(x_n)\right|} = \frac{f_X(y^2)}{\left.\left|\frac{1}{2\sqrt{x_1}}\right|\right|_{x_1=y^2}} = 2yf_X(y^2) \text{ for } y \geq 0 \blacksquare$$

Example 4.23:

Find $f_Y(y)$ given $Y= 3+(X-2)U(X-2)$, where $U(X-2)$ is the unit step function.

There is one root, $x_1 = y+2$ for $y \geq 2$. $|dy/dx| = 1$ for $y \geq 2$. Using [118]:

$$f_Y(y) = \sum_{n=1}^{1} \frac{f_X(x_n)}{\left|\frac{d}{dx}g(x_n)\right|} = \frac{f_X(y+2)}{\left.1\right|_{x_1=y+2}} = f_X(y+2) \text{ for } y \geq 2$$

However, since $g(x)$ is a constant for $x \leq 2$ there will be an impulse function at $y=3$ with a magnitude of $F_X(2)-F_X(\infty)$. See the discussion for Figure 55 and Figure 56 for reference. Notice that this method could have been used to solve example 4.14. $\blacksquare$

Before moving on to the next topic, we consider the case where $g(x)$ is continuous but $f_X(x)$ is discrete. In this case $f_Y(y)$ is also discrete. The probability of y_n is equal to the probability of x_n and y_n is related to x_n by $y_n = g(x_n)$. In the case where $g(x)$ is not monotonic[1] there will be multiple values of y. For those cases simply add the probabilities. An example will help to clarify the process.

Example 4.24:

Consider the discrete pdf in the figure below:

---

1. That is, it is not increasing only or decreasing only.

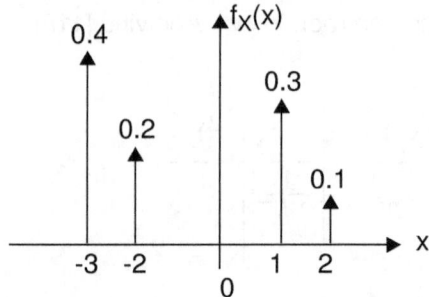

Figure 59 Discrete pdf for example 4.24

Suppose that we create a new random variable: $Y=X^2$. We can find $f_Y(y)$ as follows:

X	$Y=X^2$	P(x)
-3	9	0.4
-2	4	0.2
1	1	0.3
2	4	0.1

$f_Y(y)$ takes on the values of Y in the table above and the corresponding probabilities of x. For the case where there is more than one value of Y, the corresponding probabilities are added. The resulting pdf of Y is:

$$f_Y(y) = 0.3\delta(y-1) + 0.3\delta(y-4) + 0.4\delta(y-9)$$ ∎

4.9   Characteristic Equation

The characteristic function $\Phi_X(\omega)$ is defined as $E[e^{j\omega X}]$, where $j = \sqrt{-1}$. Thus,

[119] $\Phi_X(\omega) = E[e^{j\omega X}] = \displaystyle\int_{-\infty}^{\infty} f_X(x)e^{j\omega x}dx$

If we differentiate [119] (using Leibnitz's rule) we get

$$\frac{d}{d\omega}\int_{-\infty}^{\infty} f_X(x)e^{j\omega x}dx = \int_{-\infty}^{\infty} \frac{d}{d\omega}f_X(x)e^{j\omega x}dx = \int_{-\infty}^{\infty} jxf_X(x)e^{j\omega x}dx$$

Differentiating again,

$$\frac{d}{d\omega}\int_{-\infty}^{\infty} jxf_X(x)e^{j\omega x}dx = \int_{-\infty}^{\infty} \frac{d}{d\omega}jxf_X(x)e^{j\omega x}dx$$

$$= \int_{-\infty}^{\infty} j^2x^2f_X(x)e^{j\omega x}dx$$

Differentiating yet again,

$$\frac{d}{d\omega}\int_{-\infty}^{\infty} j^2x^2f_X(x)e^{j\omega x}dx = \int_{-\infty}^{\infty} \frac{d}{d\omega}j^2x^2f_X(x)e^{j\omega x}dx$$

$$= \int_{-\infty}^{\infty} j^3x^3f_X(x)e^{j\omega x}dx$$

A pattern can therefore be observed:

$$[120] \quad \frac{d^n}{d\omega^n}\Phi_X(\omega) = \int_{-\infty}^{\infty} j^nx^nf_X(x)e^{j\omega x}dx$$

Setting $\omega=0$ and dividing by j^n:

$$[121] \quad \frac{1}{j^n}\frac{d^n}{d\omega^n}\Phi_X(\omega)\bigg|_{\omega = 0} = \int_{-\infty}^{\infty} x^nf_X(x)dx = E[x^n]$$

Thus, the nth moment can be determined by differentiating the characteristic function of equation [119], n times and then setting ω to zero and dividing by j^n.

Example 4.25:

Determine the mean value of a random variable that is uniformly distributed from 2 to 5.

$$\Phi_X(\omega) = E[e^{j\omega X}] = \int_{-\infty}^{\infty} f_X(x)e^{j\omega x}dx = \int_{2}^{5} \frac{1}{3}e^{j\omega x}dx = \frac{1}{3j\omega}e^{j\omega x}\Big|_{2}^{5}$$

$$= \frac{1}{3j\omega}e^{j\omega 5} - \frac{1}{3j\omega}e^{j\omega 2}$$

Next take the derivative once and divide by j:

$$\frac{d}{d\omega}\Phi_X(\omega) = \frac{1}{3j^2}\frac{d}{d\omega}(\frac{1}{\omega}e^{j\omega 5} - \frac{1}{\omega}e^{j\omega 2})$$

$$= \frac{1}{-3}\left(-\frac{e^{j\omega 5}}{\omega^2} + \frac{5je^{j\omega 5}}{\omega} + \frac{e^{j\omega 2}}{\omega^2} - \frac{2je^{j\omega 2}}{\omega}\right)$$

Next we set $\omega = 0$. Since the terms are in an indeterminate form we apply L'Hôpital's rule[1] to each of the terms as follows:

$$\lim_{\omega \to 0} -\frac{e^{j\omega 5}}{\omega^2} = \lim_{\omega \to 0} -\frac{\frac{d^2}{d\omega^2}e^{j\omega 5}}{\frac{d^2}{d\omega^2}\omega^2} = \lim_{\omega \to 0} \frac{-25j^2 e^{j\omega 5}}{2} = 12.5$$

---

1. L'Hôpital's rule is: $\lim_{x \to a}\frac{u(x)}{v(x)} = \lim_{x \to a}\frac{\frac{d}{dx}u(x)}{\frac{d}{dx}v(x)} = \lim_{x \to a}\frac{\frac{d^2}{dx^2}u(x)}{\frac{d^2}{dx^2}v(x)} = ...$etc

$$\lim_{\omega \to 0} \frac{5je^{j\omega 5}}{\omega} = \lim_{\omega \to 0} \frac{\frac{d}{d\omega}5je^{j\omega 5}}{\frac{d\omega}{d\omega}} = \lim_{\omega \to 0} \frac{25j^2 e^{j\omega 5}}{1} = -25$$

$$\lim_{\omega \to 0} \frac{e^{j\omega 2}}{\omega^2} = \lim_{\omega \to 0} \frac{\frac{d^2}{d\omega^2}e^{j\omega 2}}{\frac{d^2}{d\omega^2}\omega^2} = \lim_{\omega \to 0} \frac{4j^2 e^{j\omega 2}}{2} = -2$$

$$\lim_{\omega \to 0} \frac{-2je^{j\omega 2}}{\omega} = \lim_{\omega \to 0} \frac{\frac{d}{d\omega}2je^{j\omega 2}}{\frac{d\omega}{d\omega}} = \lim_{\omega \to 0} \frac{-4j^2 e^{j\omega 2}}{1} = 4$$

Now collecting the terms gives:

$$\frac{1}{-3}(12.5 - 25 - 2 + 4) = \frac{-10.5}{-3} = 3.5$$

This is exactly what we expect and matches with [75]. ∎

4.10 Exercises

1 Which of the following are not valid cdfs and why?

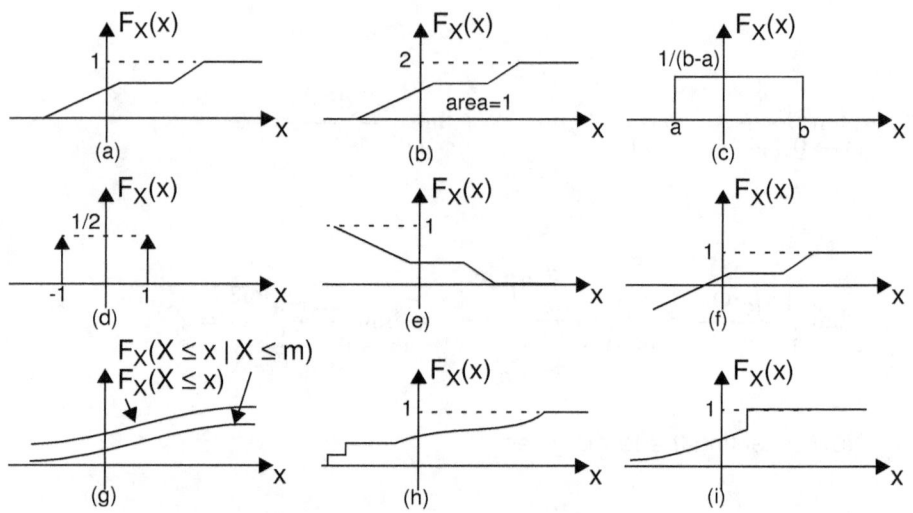

Figure 60 cdfs for Exercise 1

2 Which of the following are not valid pdfs and why?

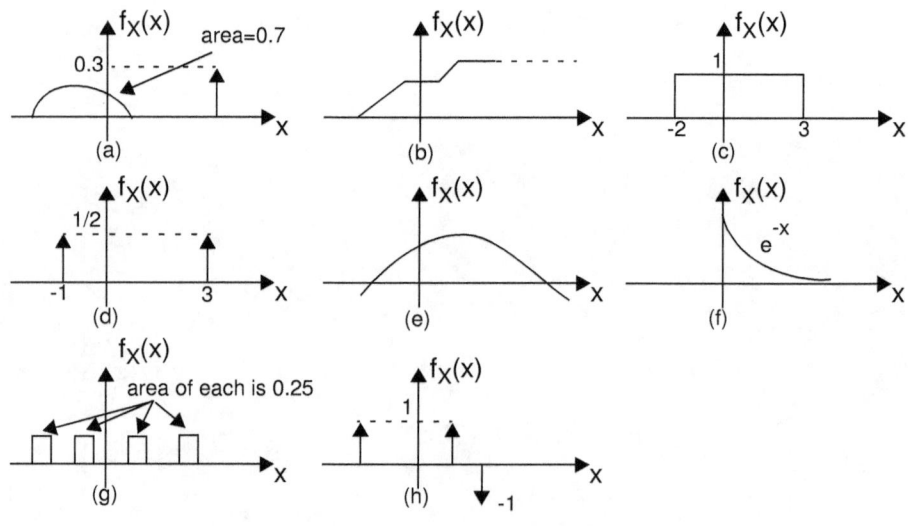

Figure 61 pdfs for Exercise 2

3 A certain electronic device has a life span that has an exponential distribution as follows: $f_T(\tau) = 1.5e^{-a\tau/\bar{\tau}}$ for $\tau \geq 0$. The mean time to failure is 36 months.

   3.1 Find a so that this is a valid pdf.

   3.2 Find the cdf.

   3.3 If τ is measured in months, find τ so that the probability of failure within τ months is 99%?

4 Find $f_Y(y)$ given $Y=e^X$.

5 Find $f_Y(y)$ given $Y=a\sin(X+\theta)$ where a>0 and X is a random variable.

6 Find $f_Y(y)$ given $Y=\tan(X)$ where a>0 and X is a random variable.

7 Find $f_Y(y)$ given $Y=aX+b$ where X is a gaussian random variable.

8 Given the circuit below, find the expected value of Y given that X is a random variable uniformly distributed from 2 to 3. Assume that the op amp remains in its linear range of operation (i.e. the output does not reach it's limit of operation).

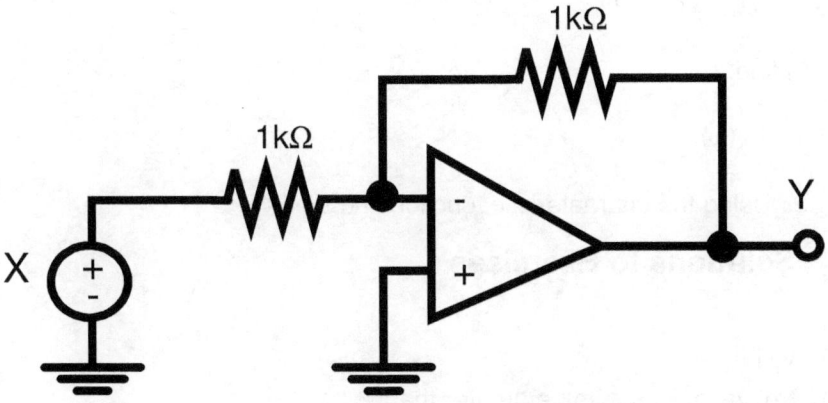

Figure 62 Circuit for Exercise 8

9 A voltage with a gaussian distribution is applied to the circuit below. Find the expected value of Vout.

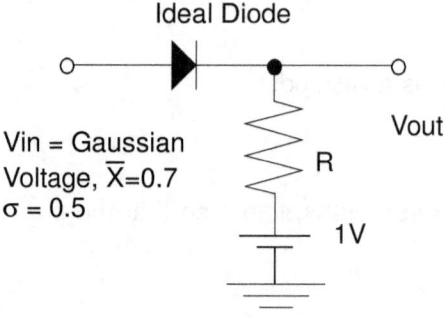

Figure 63 Circuit for Exercise 9

10 Given $f_X(x)= 0.1\delta(x+2) + 0.2\delta(x+1) + 0.5\delta(x-1) + h\delta(x-3)$ and $Y=X^2$ find:

    10.1 h such that $f_X(x)$ is a valid pdf function.

    10.2 $f_Y(y)$

11 The discrete form of the characteristic function is given by:

$$\Phi_X(\omega) = \sum_m P(X = x_m)e^{j\omega x_m}$$. If $f_X(x)= 0.1\delta(x+2) + 0.2\delta(x+1) + 0.5\delta(x-1) + 0.2\delta(x-3)$ find:

    11.1 $\Phi_X(\omega)$.

    11.2 Using the characteristic function find E[X].

4.11 Solutions to Exercises

1

    a) Valid.
    b) Not valid: Has a value greater than 1.
    c) Not valid. Not monotonically increasing.
    d) Not valid. Not monotonically increasing.
    e) Not valid. Has a negative slope.
    f) Not valid. Function goes below 0.

g) Not valid. The conditional cdf is less than the unconditional cdf.

h) Valid.

i) Valid.

2

a) Valid.

b) Not valid. Area = ∞.

c) Not valid. Area ≠ 1.

d) Valid.

e) Not valid. Function goes below 0.

f) Valid.

g) Valid.

h) Not valid. Even though area is one, the function goes below 0.

3

3.1   To be a valid pdf the distribution must have an area of 1 therefore:

$$\int_0^\infty 1.5e^{-a\frac{\tau}{36}}d\tau = 1 = -\frac{54e^{-a\frac{\tau}{36}}}{a}\Bigg|_0^\infty = -0 - \left(-\frac{54}{a}\right) = \frac{54}{a}, a=54$$

$$f_X(x) = 1.5e^{-54\frac{\tau}{36}} = 1.5e^{-1.5\tau} \ \tau \geq 0$$

3.2   Using [60]:

$$\int_{-\infty}^\tau f_T(\xi)d\xi = 1.5\int_{-\infty}^\tau e^{-1.5\xi}d\xi = 1.5\int_0^\tau e^{-1.5\xi}d\xi = -e^{-1.5\xi}\Big|_0^\tau$$

$$= 1 - e^{-1.5\tau}$$

3.3   What we are looking for is $F_T(\tau)$ = 0.99. Therefore, $0.99 = 1 - e^{-1.5\tau}$. Which reduces to $0.01 = e^{-1.5\tau}$. Taking the natural logarithm of both sides: $\ln(0.01) = -1.5\tau$. Solving for τ gives: 3.07 months.

4 $|dy/dx| = |e^x| = |y|$. There is one solution $x=\ln(y)$ for y>0. Applying [118]:

$$f_Y(y) = \sum_{n=1}^{\infty} \frac{f_X(x_n)}{\left|\frac{d}{dx}g(x_n)\right|} = \frac{f_X(\ln(y))}{y} \quad y>0$$

5 $|dy/dx| = |a\cos(x_n+\theta)|$. There are no solutions for $y>|a|$ and there are an infinite number of solutions for $y<|a|$ due to the periodic nature of the sine function. Therefore $x_n = \sin^{-1}(y/a) - \theta$, where n are all integers. Since $y=a\sin(x_n+\theta)$, we can also write $y/a = \sin(x_n+\theta)$. This would imply that a is the hypotenuse and y is the opposite side of a right triangle. Therefore the adjacent side of the triangle is $(a^2-y^2)^{1/2}$. See Figure 64. Thus, $\cos(x_n+\theta) = [(a^2-y^2)^{1/2}]/a$. Therefore, we can write $|dy/dx| = a[(a^2-y^2)^{1/2}]/a = (a^2-y^2)^{1/2}$.

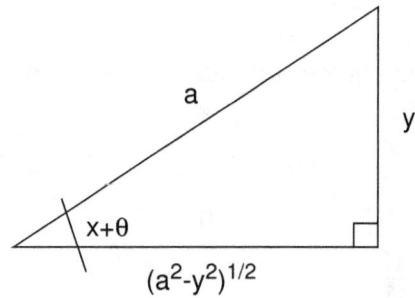

Figure 64 Sine and Cosine Determination

Applying [118]:

$$f_Y(y) = \sum_{n=-\infty}^{\infty} \frac{f_X(x_n)}{\left|\frac{d}{dx}g(x_n)\right|} = \frac{1}{\sqrt{a^2-y^2}} \sum_{n=-\infty}^{\infty} f_X(x_n) \quad |a|>y$$

6 $|dy/dx| = |1/\cos^2(x_n)|$. There are an infinite number of solutions due to the periodic nature of the tangent function. Therefore $x_n = \tan^{-1}(y)$, where n are all integers. Since $y=\tan(x)$, we can also write $y/1 = \tan(x)$. This would imply that the opposite side of a right triangle is y and the adjacent side of the triangle is 1. See Figure 65. Thus, $\cos^2(x) = (1/[(1^2+y^2)^{1/2}])^2$. Therefore, we can write $|dy/dx| = (1^2+y^2)$.

98

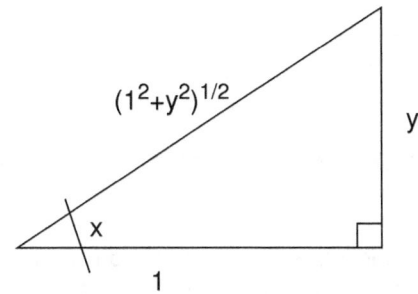

Figure 65 Tangent Determination

Applying [118]:

$$f_Y(y) = \sum_{n=-\infty}^{\infty} \frac{f_X(x_n)}{\left|\frac{d}{dx}g(x_n)\right|} = \frac{1}{1+y^2}\sum_{n=-\infty}^{\infty} f_X(x_n)$$

7 $|dy/dx| = |a|$. $x = (y-b)/a$. Applying [118]:

$$[122]\ f_Y(y) = \sum_{n=1}^{1} \frac{f_X(x_n)}{\left|\frac{d}{dx}g(x_n)\right|} = \frac{1}{\sqrt{2\pi}|a|\sigma}e^{-\frac{\left(\frac{y-b}{a}-\overline{X}\right)^2}{2\sigma^2}}$$

$$= \frac{1}{\sqrt{2\pi}|a|\sigma}e^{-\frac{(y-b-a\overline{X})^2}{2(a\sigma)^2}} = \frac{1}{\sqrt{2\pi a^2\sigma^2}}e^{-\frac{(y-(b+a\overline{X}))^2}{2(a\sigma)^2}}$$

Notice that this new pdf is also Gaussian with mean $\overline{Y}=b+a\overline{X}$ and variance $\sigma_Y^2=(a\sigma)^2$.

8 $|dy/dx| = -1$. There is one solution $x=-y$. $f_X(x)=1$ for x ranging from 2 to 3.

$$f_Y(y) = \sum_{n = -\infty}^{\infty} \frac{f_X(x_n)}{\left|\frac{d}{dx}g(x_n)\right|} = \frac{1}{|-1|} \quad -3 \le y \le -2$$

The range for y is determined by inserting the values of x into y=-x. The expected value of Y is simply [-3+(-2)]/2 = -2.5volts.

9 The input output relation is shown in Figure 66. To derive it, just note that for Vin less than one volt the diode will be reversed biased and the output will be equal to 1volt. Once Vin exceeds 1 volt, the diode becomes forward biased and Vout will equal Vin. Since the output is constant, the method used with Figure 55 and Figure 56 can be used to find the pdf of Vout. The impulse is G((1-0.7)/0.5) = G(0.6) and is located at Y=1. Using Table 2 G(0.6) = 0.7257. For Y≥1, we can use [122] with a=1 and b=0; [118] could also be used. Note that for Y≥1 Y=X so $f_Y(y)=f_X(y)$. The resulting pdf is shown in Figure 67.

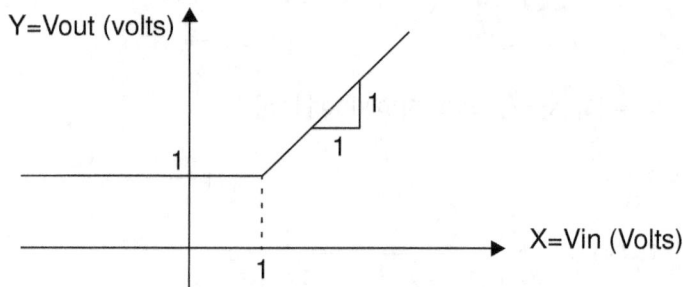

Figure 66 Vout vs. Vin

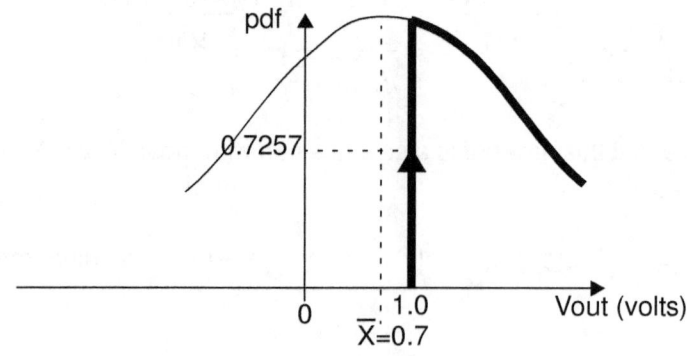

Figure 67 Vout vs. Vin

Note that [110] cannot be used to calculate the mean in this case. The reason is, [110] was derived under the assumption that the conditioning value is equal to the mean of the unconditional pdf. So in this case we have to rely on [64] to find the mean. Also compare Figure 67 with Figure 53. Therefore:

$$\overline{Vout} = \int_{-\infty}^{\infty} x f_X(x) dx = \int_{-\infty}^{\infty} 0.7257 x \delta(x-1) dx + \int_{1}^{\infty} x \frac{e^{-\frac{(x-0.7)^2}{2(0.5^2)}}}{\sqrt{2\pi} 0.5} dx$$

If you got this far consider the problem solved. However, if you have a math program that can perform the integration the answer should be 1.084volts.

10

10.1 Using [59]: $\int_{-\infty}^{\infty} f_X(x) dx = 1$. Therefore,

$$\int_{-\infty}^{\infty} (0.1\delta(x+2) + 0.2\delta(x+1) + 0.5\delta(x-1) + h\delta(x-3)) dx$$

$$= 0.1 + 0.2 + 0.5 + h = 0.8 + h = 1 \rightarrow h = 0.2$$

10.2 $f_X(x) = 0.1\delta(x+2) + 0.2\delta(x+1) + 0.5\delta(x-1) + 0.2\delta(x-3)$

X	Y=X²	P(x)
-2	4	0.1
-1	1	0.2
1	1	0.5
3	9	0.2

$f_Y(y)$ takes on the values of Y in the table above and the corresponding probabilities of x. For the case where there is more than one value of Y, the corresponding probabilities are added. The resulting pdf of Y is:

$$f_Y(y) = 0.7\delta(y-1) + 0.1\delta(y-4) + 0.2\delta(y-9)$$

11

11.1 $$\Phi(\omega) = \sum_m P(X = x_m) e^{j\omega x_m}$$

$$= 0.1e^{-j\omega 2} + 0.2e^{-j\omega 1} + 0.5e^{j\omega 1} + 0.2e^{j\omega 3}$$

11.2 Using [121]:

$$\frac{1}{j}\frac{d}{d\omega}\Phi(\omega)\bigg|_{\omega=0} = \frac{-0.2je^{-j\omega 2} - 0.2je^{-j\omega 1} + 0.5je^{j\omega 1} + 0.6je^{j\omega 3}}{j}\bigg|_{\omega=0}$$

$$= -0.2 - 0.2 + 0.5 + 0.6 = 0.7$$

This is the same result that would have been obtained if [65] were used.

Chapter 5 Multiple Random Variables

In experiments, there are often multiple random variables that need to be considered in the sample space. In this chapter we develop the tools to allow us to work with multiple random variables.

5.1   Joint cdf

Recall from section 2.3 that the joint probability of two events is defined as $P(A \cap B)$. We take this definition and define the joint cdf of two random variables, X and Y, as:

[123] $F_{XY}(x, y) = P(\{X \leq x\} \cap \{Y \leq y\})$ for the continuous case

[124] $F_{XY}(x, y) = \sum_i \sum_j P(X = x_i \cap Y = y_j) U(x - x_i) U(y - y_j)$ for the discrete case.

Since there are only two random variables involved we call these cdfs bivariate cdfs.

Example 5.1:

In this example, we will show how to apply [124]. Assume that there are two outputs of a digital circuit. The 1st output is 1-bit and the 2nd output is 2-bits. Let us define the random variables X and Y to be the number of 1s in the outputs. Remember from Chapter 2, that a random variable is the mapping the experimental outcome to the real line. We do the same here for each random variable and assign probabilities as follows:

OUT1 = 0	$\rightarrow$	X = 0	P(0)=1/2
OUT1 = 1	$\rightarrow$	X = 1	P(1)=1/2
OUT2 = 00	$\rightarrow$	Y = 0	P(0)=1/4
OUT2 = 01 or 10	$\rightarrow$	Y = 1	P(1)=1/2
OUT2 = 11	$\rightarrow$	Y = 2	P(2)=1/4

Note that the mapping chosen was arbitrary. Now $F_{XY}(x,y)$ can be calculated. Assuming that the two sources are independent[1] the cdf can be tabulated by first calculating the joint probabilities which are basically the probability masses for the joint events. The number of (X,Y) pairs is seen to be 6: (0,0), (0,1), (0,2), (1,0), (1,1), and (1,2). Therefore, we will need to calculate 6 probabilities.

$P(X=0 \cap Y=0) = 1/2 * 1/4 = 1/8$

$P(X=0 \cap Y=1) = 1/2 * 1/2 = 1/4$

$P(X=0 \cap Y=2) = 1/2 * 1/4 = 1/8$

$P(X=1 \cap Y=0) = 1/2 * 1/4 = 1/8$

$P(X=1 \cap Y=1) = 1/2 * 1/2 = 1/4$

$P(X=1 \cap Y=2) = 1/2 * 1/4 = 1/8$

Then the cdf is calculated by inserting the joint probabilities into [124] as follows:

[125] $F_{XY}(x,y) = P(X=0 \cap Y=0)U(x-0)U(y-0) + P(X=0 \cap Y=1)U(x-0)U(y-1) + P(X=0 \cap Y=2)U(x-0)U(y-2) + P(X=1 \cap Y=0)U(x-1)U(y-0) + P(X=1 \cap Y=1)U(x-1)U(y-1) + P(X=1 \cap Y=2)U(x-1)U(y-2) = 1/8U(x-0)U(y-0) + 1/4U(x-0)U(y-1) + 1/8U(x-0)U(y-2) + 1/8U(x-1)U(y-0) + 1/4U(x-1)U(y-1) + 1/8U(x-1)U(y-2)$

As it was for the discrete case with a single variable, the cdf is a step function. [125] is then plotted as a three dimensional plot of the cdf as shown below in Figure 68.

---

1. Independence will be covered in a subsequent section. Assume for now that we can multiply the probabilities together and that X and Y are independent; see example 5.6. This is safe because the outputs of the logic circuits do not depend upon their previous outputs or each other.

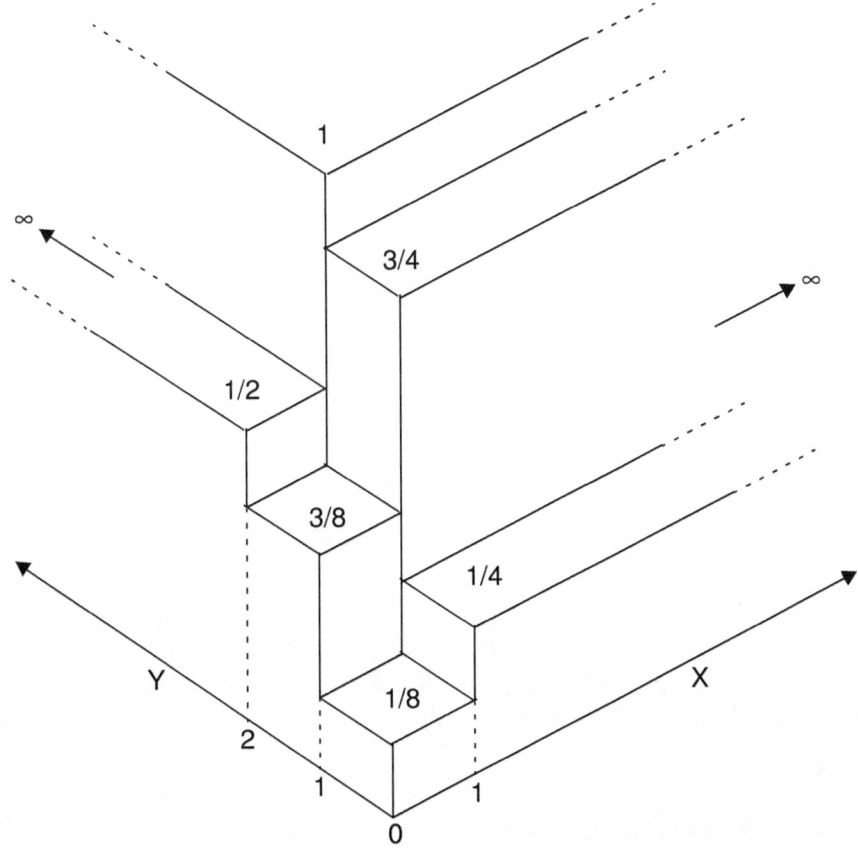

Figure 68 $F_{XY}(x,y)$ for example 5.1

■

We can directly expand [123] to accommodate N random variables as follows to make an n-variate cdf:

$$[126] \quad F_{X_1 X_2 \dots X_N}(x_1, x_2 \dots x_N) = P(\{X_1 \le x_1\} \cap \{X_2 \le x_2\} \cap \dots \{X_N \le x_N\})$$

[124] can also be expanded in a similar manner. Multiple mixed random variables are beyond the scope of this text and, thus, we will only consider cases where the random variables are all discrete or all continuous.

5.2   Properties of the Bivariate cdf

The properties of the continuous bivariate (or joint) cdf are listed below:

[127] $F_{XY}(-\infty, -\infty) = 0$

[128] $F_{XY}(x, -\infty) = 0$

[129] $F_{XY}(-\infty, y) = 0$

[130] $0 \le F_{XY}(x, y) \le 1$

[131] $P(x_1 < X \le x_2 \cap y_1 < Y \le y_2) = F_{XY}(x_2, y_2) + F_{XY}(x_1, y_1)$
$\qquad -F_{XY}(x_1, y_2) - F_{XY}(x_2, y_1)$

[132] $F_{XY}(x, \infty) = F_X(x)$

[133] $F_{XY}(\infty, y) = F_Y(y)$

[134] $F_{XY}(x, y)$ is a monotonically increasing function of x and y

[135] $F_{XY}(\infty, \infty) = 1$

With the exception of [132] and [133], the above properties can be used as tests to verify that a joint cdf is valid. [132] and [133] are called marginal cdfs.

Example 5.2:

Prove properties [127] - [133].

To prove property [127] note that $F_{XY}(-\infty,-\infty) = P(\{X \le -\infty\} \cap \{Y \le -\infty\})$. Note that the set $\{X \le -\infty\} = \{Y \le -\infty\} = \varnothing$. Thus, $F_{XY}(-\infty,-\infty) = P(\varnothing \cap \varnothing)=0$[1].

To prove property [128] note that $F_{XY}(x,-\infty) = P(\{X \le x\} \cap \{Y \le -\infty\})$. Note that the set $\{Y \le -\infty\} = \varnothing$. Thus, $F_{XY}(x,-\infty) = P(\{X \le x\} \cap \varnothing)=0$.

To prove property [129] note that $F_{XY}(-\infty,y) = P(\{X \le -\infty\} \cap \{Y \le y\})$. Note that the set $\{X \le -\infty\} = \varnothing$. Thus, $F_{XY}(-\infty,y) = P(\varnothing \cap \{Y \le y\})=0$.

To prove property [130] note that $F_{XY}(x,y)$ is a probability and as such must satisfy the axioms of probability, notably [20] on page 20

---

1. See equation [6] on page 10.

To prove property [131] note that $\{x_1 < X \le x_2 \cap Y \le y_2\} = \{x_1 < X \le x_2 \cap Y \le y_1\} \cup \{x_1 < X \le x_2 \cap y_1 < Y \le y_2\}$. Also note that the right two terms are disjoint, so we can write:

$$P(\{x_1 < X \le x_2 \cap Y \le y_2\}) = P(\{x_1 < X \le x_2 \cap Y \le y_1\}) + P(\{x_1 < X \le x_2 \cap y_1 < Y \le y_2\})$$

Rearranging the terms gives:

[136] $P(\{x_1 < X \le x_2 \cap y_1 < Y \le y_2\}) = P(\{x_1 < X \le x_2 \cap Y \le y_2\}) - P(\{x_1 < X \le x_2 \cap Y \le y_1\})$.

Now note that $\{X \le x_2 \cap Y \le y_2\} = \{X \le x_1 \cap Y \le y_2\} \cup \{x_1 < X \le x_2 \cap Y \le y_2\}$ and that the right two terms are disjoint. Therefore, we can write:

$$P(\{X \le x_2 \cap Y \le y_2\}) = P(\{X \le x_1 \cap Y \le y_2\}) + P(\{x_1 < X \le x_2 \cap Y \le y_2\})$$

Rearranging the terms gives:

[137] $P(\{x_1 < X \le x_2 \cap Y \le y_2\}) = P(\{X \le x_2 \cap Y \le y_2\}) - P(\{X \le x_1 \cap Y \le y_2\})$

Similarly we can write:

[138] $P(\{x_1 < X \le x_2 \cap Y \le y_1\}) = P(\{X \le x_2 \cap Y \le y_1\}) - P(\{X \le x_1 \cap Y \le y_1\})$

Inserting [137] and [138] into [136] results in:

$$P(\{x_1 < X \le x_2 \cap y_1 < Y \le y_2\}) = P(\{X \le x_2 \cap Y \le y_2\}) - P(\{X \le x_1 \cap Y \le y_2\}) - P(\{X \le x_2 \cap Y \le y_1\}) + P(\{X \le x_1 \cap Y \le y_1\})$$

This can be written as $P(\{x_1 < X \le x_2 \cap y_1 < Y \le y_2\}) = F_{XY}(x_2,y_2) - F_{XY}(x_1,y_2) - F_{XY}(x_2,y_1) + F_{XY}(x_1,y_1)$, which is the same as [131].

To prove property [132] note that $F_{XY}(x,\infty) = P(\{X \le x\} \cap \{Y \le \infty\})$. Note that the set $\{Y \le \infty\} = \Omega$. Thus, $F_{XY}(x,\infty) = P(\{X \le x\} \cap \Omega) = P(\{X \le x\})$ which is $F_X(x)$.

To prove property [133] note that $F_{XY}(\infty,y) = P(\{X \le \infty\} \cap \{Y \le y\})$. Note that the set $\{X \le \infty\} = \Omega$. Thus, $F_{XY}(\infty,y) = P(\Omega \cap \{Y \le y\}) = P(\{Y \le y\})$ which is $F_Y(y)$. ■

Example 5.3:

Referring back to example 5.1 find $P(\{0 < X \leq 1 \cap 0 < Y \leq 1\})$. This can be found from [131] as: $P(\{x_1 < X \leq x_2 \cap y_1 < Y \leq y_2\}) = F_{XY}(1,1) - F_{XY}(0,1) - F_{XY}(1,0) + F_{XY}(0,0) = 3/4 - 3/8 - 1/4 + 1/8 = 1/4$. This result may not be obvious from examining Figure 68. However another method can be used. We restate [125] below:

$F_{XY}(x,y) = 1/8U(x-0)U(y-0) + 1/4U(x-0)U(y-1) + 1/8U(x-0)U(y-2) + 1/8U(x-1)U(y-0) + 1/4U(x-1)U(y-1) + 1/8U(x-1)U(y-2)$

Next the terms that do not meet the requirement of the event $0 < X \leq 1$ are removed. This corresponds to the first three terms because they only have meaning at x=0. The last three terms are left because they have meaning at x=1 which meets the requirement $X \leq 1$. Now let's look at the other requirement $0 < Y \leq 1$. The 4th term does not meet the requirement for $0 < y$, so it is removed. The 6th term does not meet the requirement for $y \leq 1$ so it is removed. The remaining term, $1/4U(x-1)U(y-1)$ meets both requirements for $0 < X \leq 1$ and $0 < Y \leq 1$, therefore it remains and the answer is 1/4 as was calculated above using [131]. ∎

5.3   Bi-variate pdf and pmf

The bi-variate pdf is found from the double partial differentiation below:

$$[139] \quad f_{XY}(x, y) = \frac{\partial^2}{\partial x \partial y} F_{XY}(x, y)$$

This can be expanded to handle N random variables as shown below:

$$[140] \quad f_{X_1 X_2 \ldots X_N}(x_1, x_2 \ldots x_N) = \frac{\partial^N}{\partial x_1 \partial x_2 \ldots \partial x_N} F_{X_1 X_2 \ldots X_N}(x_1, x_2 \ldots x_N)$$

[140] is known as an n-variate pdf. By inserting [124] into [139] we get:

$$[141] \quad f_{XY}(x, y) = \sum_i \sum_j P(X = x_i \cap Y = y_j) \delta(x - x_i) \delta(y - y_j) \quad \text{which is a}$$

bi-variate probability mass function for two discrete random variables.

5.4   Properties of the Bi-variate pdf and pmf

The properties of the Bi-Variate pdf are listed below. Note the similarity to the pdf properties for a single random variable.

[142] $f_{XY}(x, y) \geq 0$

[143] $\displaystyle\int_{-\infty}^{\infty} \int_{-\infty}^{\infty} f_{XY}(x, y) dx dy = 1$ [1]

[144] $F_{XY}(x, y) = \displaystyle\int_{-\infty}^{y} \int_{-\infty}^{x} f_{XY}(\xi_1, \xi_2) d\xi_1 d\xi_2$

[145] $F_X(x) = \displaystyle\int_{-\infty}^{x} \int_{-\infty}^{\infty} f_{XY}(\xi_1, \xi_2) d\xi_2 d\xi_1$

[146] $F_Y(y) = \displaystyle\int_{-\infty}^{y} \int_{-\infty}^{\infty} f_{XY}(\xi_1, \xi_2) d\xi_1 d\xi_2$

[147] $P(x_1 < X \leq x_2 \cap y_1 < Y \leq y_2) = \displaystyle\int_{y_1}^{y_2} \int_{x_1}^{x_2} f_{XY}(x, y) dx dy$

[148] $f_X(x) = \displaystyle\int_{-\infty}^{\infty} f_{XY}(x, y) dy$

[149] $f_Y(y) = \displaystyle\int_{-\infty}^{\infty} f_{XY}(x, y) dx$

[150] $F_{X_1 X_2 \cdots X_N}(x_1, x_2 \cdots x_N)$

$= \displaystyle\int_{-\infty}^{x_N} \cdots \int_{-\infty}^{x_2} \int_{-\infty}^{x_1} f_{X_1 X_2 \cdots X_N}(\xi_1, \xi_2 \cdots \xi_N) d\xi_1 d\xi_2 \cdots d\xi_N$

---

1. The interpretation of this is that the volume under the joint pdf is equal to 1.

Example 5.4:

A joint pdf is given by $f_{XY}(x,y) = x^2 + ky$ for $0 \le x \le 1, 0 \le y \le 1$. Find:

a) k, to make the pdf valid
b) the cdf
c) $P(x \le 1/2 \cap y > 1/2)$

a)   Using [143], $\displaystyle\int_{-\infty}^{\infty}\int_{-\infty}^{\infty} f_{XY}(x,y)dxdy = 1$ and integrating,

$$\int_0^1\int_0^1 (x^2 + ky)dxdy = \int_0^1 \left(\frac{x^3}{3} + kxy\Big|_0^1\right)dx = \int_0^1 \left(\frac{1}{3} + ky\right)dy$$

$$\frac{1}{3}y + k\frac{y^2}{2}\Big|_0^1 = \frac{1}{3} + \frac{k}{2} = 1 \qquad k \to \frac{4}{3}$$

Which means that k needs to be 4 to make the pdf valid.

b)   Using [144], $\displaystyle F_{XY}(x,y) = \int_{-\infty}^{y}\int_{-\infty}^{x} f_{XY}(\xi_1, \xi_2)d\xi_1 d\xi_2$ and

integrating

$$F_{XY}(x,y) = \int_0^y\int_0^x \left(\xi_1^2 + \frac{4}{3}\xi_2\right)d\xi_1 d\xi_2 = \int_0^y \left(\frac{\xi_1^3}{3} + \frac{4}{3}\xi_1\xi_2\Big|_0^x\right)d\xi_2$$

$$= \int_0^y \left(\frac{x^3}{3} + \frac{4}{3}x\xi_2\right)d\xi_2 = \frac{x^3}{3}\xi_2 + \frac{4}{3}x\frac{\xi_2^2}{2}\Big|_0^y = \frac{x^3 y}{3} + \frac{2xy^2}{3}$$

As a spot check we use [139], $f_{XY}(x,y) = \dfrac{\partial^2}{\partial x \partial y}F_{XY}(x,y)$ to verify
that we end up with the original pdf:

$$f_{XY}(x,y) = \frac{\partial^2}{\partial x \partial y}F_{XY}(x,y) = x^2 + \frac{4}{3}y, \text{ which is equal to the original}$$

pdf.

110

c) To find P(x ≤ 1/2 ∩ y > 1/2), [147] can be called into action:

$$P\left(-\infty < X \le \frac{1}{2} \cap \frac{1}{2} < Y \le \infty\right) = \int_{\frac{1}{2}}^{1}\int_{0}^{\frac{1}{2}}\left(x^2 + \frac{4}{3}y\right)dxdy = \int_{\frac{1}{2}}^{1}\left(\frac{x^3}{3} + \frac{4}{3}yx\bigg|_{0}^{\frac{1}{2}}\right)dy$$

$$= \int_{\frac{1}{2}}^{1}\frac{2}{3}y + \frac{1}{24}dy = \frac{2}{6}y^2 + \frac{1}{24}y\bigg|_{\frac{1}{2}}^{1} = \frac{13}{48} \quad \blacksquare$$

The properties of the discrete bivariate pmf are:

[151] $$\sum_{i=-\infty}^{\infty}\sum_{j=-\infty}^{\infty} P(X = x_i \cap Y = y_j) = 1$$

[152] $0 \le P(X = x_i \cap Y = y_j) \le 1$

[153] $$P(X = x_i) = \sum_{j} P(X = x_i \cap Y = y_j)\delta(x - x_i)$$

[154] $$P(Y = y_j) = \sum_{i} P(X = x_i \cap Y = y_j)\delta(y - y_j)$$

[153] and [154] are called the marginal pmfs.

Example 5.5:

We next demonstrate how to use [153] and [154] to calculate the marginal pmfs for example 5.1. Starting with [153]:

$f_X(x)$ = P(X=0 ∩ Y=0)δ(x-0) + P(X=0 ∩ Y=1)δ(x-0) + P(X=0 ∩ Y=2)δ(x-0) + P(X=1 ∩ Y=0)δ(x-1) + P(X=1 ∩ Y=1)δ(x-1) + P(X=1 ∩ Y=2)δ(x-1) = (1/8 + 1/4 + 1/8)δ(x-0) + (1/8 + 1/4 + 1/8)δ(x-1) = 1/2δ(x-0) + 1/2δ(x-1) which is the pmf for X.

111

Next, using [154]:

$f_Y(y)$ = P(X=0 ∩ Y=0)δ(y-0) + P(X=0 ∩ Y=1)δ(y-1) + P(X=0 ∩ Y=2)δ(y-2) + P(X=1 ∩ Y=0)δ(y-0) + P(X=1 ∩ Y=1)δ(y-1) + P(X=1 ∩ Y=2)δ(y-2) = (1/8 + 1/8)δ(y-0) + (1/4 + 1/4)δ(y-1) + (1/8 + 1/8)δ(y-2) = 1/4δ(y-0) + 1/2δ(y-1) + 1/4δ(y-2) which is the pmf for Y.

Note that we needed to apply [153] twice to take into consideration the two values of X. Likewise we needed to apply [154] three times to take into consideration the three values of Y. ∎

5.5　Independence

Independence was defined by [41] on page 28 as P(A ∩ B) = P(A)P(B). This definition is applied to [123] and [124] with the following results:

[155] $F_{XY}(x, y) = P(\{X \le x\} \cap \{Y \le y\}) = P(\{X \le x\})P(\{Y \le y\})$

$= F_X(x)F_Y(y)$ for continuous X and Y.

[156] $F_{XY}(x, y) = \sum_i \sum_j P(X = x_i \cap Y = y_j)U(x - x_i)U(y - y_j)$

$= \sum_i \sum_j P(X = x_i)P(Y = y_j)U(x - x_i)U(y - y_j)$ for discrete X and Y.

Note that in [155] and [156] the marginal cdfs are used. Inserting [155] and [156] into [139] gives the pdfs and pmfs for when X and Y are independent:

[157] $f_{XY}(x, y) = \frac{\partial^2}{\partial x \partial y}F_{XY}(x, y) = \frac{\partial^2}{\partial x \partial y}F_X(x)F_Y(y) = f_X(x)f_Y(y)$ for continuous X and Y.

$$[158] \quad f_{XY}(x, y) = \frac{\partial^2}{\partial x \partial y} F_{XY}(x, y)$$

$$= \frac{\partial^2}{\partial x \partial y} \sum_i \sum_j P(X = x_i) P(Y = y_j) U(x - x_i) U(y - y_j)$$

$$= \sum_i \sum_j P(X = x_i) P(Y = y_j) \delta(x - x_i) \delta(y - y_j)$$

As indicated in section 2.7, it is not enough that pair-wise independence be shown to prove independence. The modified version of [42]-[44] below must be met:

$$[159] \quad F_{A_i A_j}(a_i, a_j) = F_{A_i}(a_i) F_{A_j} \quad , \text{ for all i and j}$$

$$[160] \quad F_{A_i A_j A_k}(a_i, a_j, a_k) = F_{A_i}(a_i) F_{A_j}(a_j) F_{A_k}(a_k) \quad , \text{ for all i, j, and k}$$

...

$$[161] \quad F_{A_i A_j \ldots A_N}(a_i, a_j, \ldots a_N) = F_{A_i}(a_i) F_{A_j}(a_j) \ldots F_{A_N}(a_N) \quad \text{for all i through}$$

N

Next are some examples to illustrate the concepts.

Example 5.6:

Returning to example 5.1, we wish to determine if our initial assumption that X and Y are independent holds true. To prove this, we must demonstrate that [156] holds true. We can see that the sample space $\Omega = \{0/00; 0/01; 0/10; 0/11; 1/00; 1/01; 1/10; 1/11\}$. The slash is used to separate OUT1 and OUT2. Since there are 8 elements and each element is equally likely we can write the middle term of [156] as:

$1/8U(x-0)U(y-0) + (1/8 + 1/8)U(x-0)U(y-1)^1 + 1/8U(x-0)U(y-2) + 1/8U(x-1)U(y-0) + (1/8 + 1/8)U(x-1)U(y-1)^2 + 1/8U(x-1)U(y-2)$

---

1. 1/8+1/8 comes from the fact that there are two cases where X=0 and Y=1 namely 0/01 and 0/10.
2. 1/8+1/8 comes from the fact that there are two cases where X=1 and Y=1 namely 1/01 and 1/10.

= 1/8U(x-0)U(y-0) + 1/4U(x-0)U(y-1) + 1/8U(x-0)U(y-2) + 1/8U(x-1)U(y-0) + 1/4U(x-1)U(y-1) + 1/8U(x-1)U(y-2).

Now let's see if the last term of [156] matches:

Using the joint probabilities:

(1/2)(1/4)U(x-0)U(y-0) + (1/2)(1/2)U(x-0)U(y-1) + (1/2)(1/4)U(x-0)U(y-2) + (1/2)(1/4)U(x-1)U(y-0) + (1/2)(1/2)U(x-1)U(y-1) + (1/2)(1/4)U(x-1)U(y-2)

= 1/8U(x-0)U(y-0) + 1/4U(x-0)U(y-1) + 1/8U(x-0)U(y-2) + 1/8U(x-1)U(y-0) + 1/4U(x-1)U(y-1) + 1/8U(x-1)U(y-2).

The results match and, therefore, our previous assumption about X (OUT1) and Y (OUT2) being independent was correct. ∎

Example 5.7:

Given the marginal pdfs for two independent random variables X and Y:

$f_X(x) = e^{-x}$ $0 < x < \infty$ & $f_Y(y) = 0.5e^{-|y-1|}$ $-\infty < y < \infty$, What is $P(x > 1 \cap y > 0)$?

We can use [147]. Note that because X and Y are independent, $f_{XY}(x,y) = f_X(x)f_Y(y)$ as indicated by [155]. Thus,

$$P(1 < X \le \infty, 0 < Y \le \infty) = \int_0^\infty \int_1^\infty 0.5e^{-x}e^{-|y-1|}dxdy$$

$$= 0.5\int_0^\infty e^{-|y-1|}\left(\int_1^\infty e^{-x}dx\right)dy$$

$$= 0.5\int_0^\infty e^{-|y-1|}\left(-e^{-x}\Big|_1^\infty\right)dy = 0.5e^{-1}\int_0^\infty e^{-|y-1|}dy$$

$$= 0.5e^{-1} \left(\int_0^1 e^{-(1-y)}dy + \int_1^\infty e^{-(y-1)}dy \right) = 0.5e^{-1}(2 - e^{-1})$$

$$= 0.30021 \blacksquare$$

Example 5.8:

For the two cases below, determine if X and Y are independent.

a) $f_{XY}(x, y) = k\dfrac{x}{e^y}$ $0 \le x \le 1,\ 0 \le y \le 1$

b) $f_{XY}(x, y) = k(xy + y)$ $0 \le x \le 1,\ 0 \le y \le 1$

a)   If the product of the marginal cdfs is equal to the joint cdf then X and Y are independent. First [143] is used to determine k:

$$\int_0^1 \int_0^1 k\frac{x}{e^y}dxdy = k\frac{x^2}{2}\bigg|_0^1 \int_0^1 \frac{1}{e^y}dy = \frac{k}{2}\ln(e^y)\bigg|_0^1 = \frac{k\ln(e^1)}{2} = \frac{k}{2} = 1$$

This means that k=2. So the joint pdf is:

$$f_{XY}(x, y) = \frac{2x}{e^y}\ \ 0 \le x \le 1,\ 0 \le y \le 1$$

Using [148] to determine the first marginal probability:

$$f_X(x) = \int_0^1 2\frac{x}{e^y}dy = 2x\ln(e^y)\bigg|_0^1 = 2x[\ln(e^1) - \ln(e^0)] = 2x$$

Next using [149] to determine the second marginal probability:

$$f_Y(y) = \int_0^1 2\frac{x}{e^y}dx = \frac{2}{e^y}\frac{x^2}{2}\bigg|_0^1 = \frac{1}{e^y}$$

Multiplying the two marginal pdfs together gives:

$$f_X(x)f_Y(y) = \frac{2x}{e^y} = f_{XY}(x, y)$$

Since the product of the marginal pdfs matches the joint cdf, X and Y are independent.

b)   If the product of the marginal cdfs is equal to the joint cdf then X and Y are independent. First [143] is used to determine k:

$$\int_0^1 \int_0^1 k(xy + y)dxdy = k\int_0^1 \left(\frac{x^2 y}{2}\Big|_0^1 + yx\Big|_0^1\right)dy = k\int_0^1 \left(\frac{y}{2} + y\right)dy$$

$$= k\left(\frac{y^2}{4}\Big|_0^1 + \frac{y^2}{2}\Big|_0^1\right) = \frac{3k}{4} = 1$$

k is then found to be 4/3.

Using [148] to determine the first marginal probability:

$$f_X(x) = \int_0^1 \frac{4}{3}(xy + y)dy = \frac{4}{3}\left(x\frac{y^2}{2}\Big|_0^1 + \frac{y^2}{2}\Big|_0^1\right) = \frac{4}{3}\left(\frac{x}{2} + \frac{1}{2}\right) = \frac{2}{3}x + \frac{2}{3}$$

Next using [149] to determine the second marginal probability:

$$f_Y(y) = \int_0^1 \frac{4}{3}(xy + y)dx = \frac{4}{3}\left(\frac{x^2}{2}y\Big|_0^1 + xy\Big|_0^1\right) = \frac{4}{3}\left(\frac{y}{2} + y\right) = \frac{2}{3}y + \frac{4}{3}$$

Multiplying the marginal pdfs gives:

$$f_X(x)f_Y(y) = \left(\frac{2}{3}x + \frac{2}{3}\right)\left(\frac{2}{3}y + \frac{4}{3}\right) \neq f_{XY}(x, y)$$

Since the product of the marginal pdfs does not equal the joint pdf, X and Y are concluded to not be statistically independent. ∎

5.6   Joint Expectation, Moments, and Variance

Joint expectation, moments and variance can be defined for multiple random variables in a similar manner as was done for single random variables in section 4.5. The expectation of the function of two continuous random variables is defined as:

$$[162]\ \ E[h(X, Y)] = \int_{-\infty}^{\infty} \int_{-\infty}^{\infty} h(x, y) f_{XY}(x, y) dx dy$$

E[XY] is called the correlation of X and Y and is discussed in more detail in section 5.9.

For N random variables [162] can be expanded as follows:

$$[163]\ \ E[h(X_1, ..., X_N)]$$

$$\int_{-\infty}^{\infty} ... \int_{-\infty}^{\infty} h(x_1, ..., x_N) f_{X_1, ..., X_N}(x_1, ..., x_N) dx_1 ... dx_N$$

The moments of two random variables X and Y is defined as:

$$[164]\ \ E[X^n Y^m] = \int_{-\infty}^{\infty} \int_{-\infty}^{\infty} x^n y^m f_{XY}(x, y) dx dy$$

where n+m is called the order of the moment. [164] can be expanded to accommodate N random variables as follows:

$$[165]\ \ E\left[X_1^{n_1}, ..., X_N^{n_N}\right]$$

$$= \int_{-\infty}^{\infty} ... \int_{-\infty}^{\infty} x_1^{n_1}, ..., x_N^{n_N} f_{X_1, ..., X_N}(x_1, ..., x_N) dx_1 ... dx_N$$

The central moment for two random variables is defined as:

$$[166]\ \ E[(X - \bar{X})^n (Y - \bar{Y})^m] = \int_{-\infty}^{\infty} \int_{-\infty}^{\infty} (x - \bar{x})^n (y - \bar{y})^m f_{XY}(x, y) dx dy$$

As with the cases above, this can be expanded to N random variables:

[167] $E\left[(X_1 - \overline{X_1})^{n_1}...(X_N - \overline{X_N})^{n_N}\right]$

$$= \int_{-\infty}^{\infty} \int_{-\infty}^{\infty} (X_1 - \overline{X_1})^{n_1}...(X_N - \overline{X_N})^{n_N} f_{X_1...X_N}(x_1...x_N) dx_1...dx_N$$

For discrete random variables the moments are defined as:

[168] $E[X^n Y^m] = \sum_{x_i} \sum_{y_j} x_i^n y_j^m P(X = x_i \cap Y = y_j)$

Note that if n=1 and m=0 in [168] the result is the mean of X. Likewise if n=0 and m=1 the result is the mean of Y.

Example 5.9:

Referring back to example 5.1, what is $E[XY^2]$?

Using [168], we can write:

$E[X^1 Y^2] = 0^1 0^2 P(X = 0 \cap Y = 0) + 0^1 1^2 P(X = 0 \cap Y = 1)$

$+ 0^1 2^2 P(X = 0 \cap Y = 2) + 1^1 0^2 P(X = 1 \cap Y = 0)$

$+ 1^1 1^2 P(X = 1 \cap Y = 1) + 1^1 2^2 P(X = 1 \cap Y = 2)$

$= \frac{1}{4} + 4\left(\frac{1}{8}\right) = \frac{3}{4}$ ∎

Example 5.10:

Given the joint pdf in the figure below what is $E[XY^2]$?

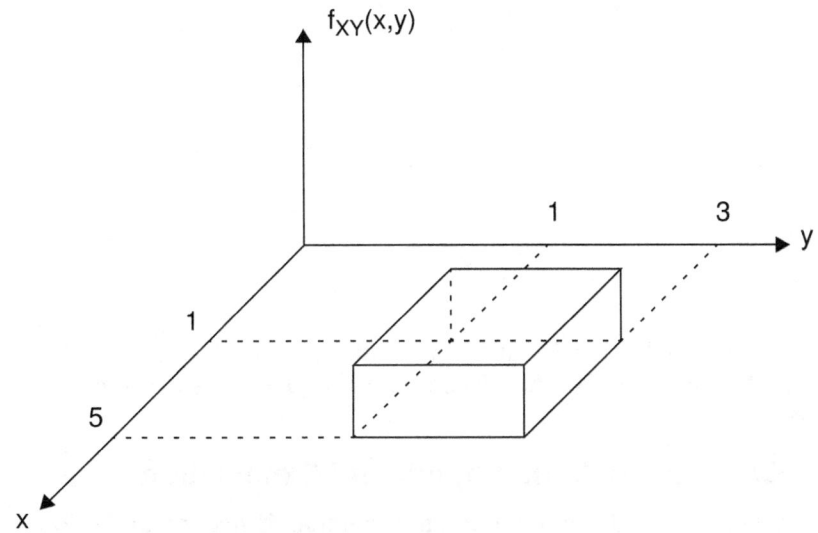

Figure 69 pdf for example 5.10

[164] can be used. Next $f_{XY}(x,y)$ is $1/[(5-1)(3-1)]$. Next [164] is used:

$$E[XY^2] = \int_1^3 \int_1^5 \frac{1}{8}xy^2 dx dy = \frac{1}{16}\int_1^3 x^2 y^2 \Big|_1^5 dy = \frac{24}{16}\int_1^3 y^2 dy$$

$$= \frac{3}{2}\int_1^3 y^2 dy = \frac{3}{6}y^3 \Big|_1^3 = \frac{1}{2}(27-1) = 13 \quad \blacksquare$$

Example 5.11:

What is the expected value of the sum of two random variables?

$$E[X+Y] = \int_{-\infty}^{\infty} \int_{-\infty}^{\infty} (x+y)f_{XY}(x, y)dx dy$$

$$= \int_{-\infty}^{\infty} \int_{-\infty}^{\infty} xf_{XY}(x, y)dy dx + \int_{-\infty}^{\infty} \int_{-\infty}^{\infty} yf_{XY}(x, y)dx dy$$

$$= \int_{-\infty}^{\infty} x \int_{-\infty}^{\infty} f_{XY}(x, y) dy dx + \int_{-\infty}^{\infty} y \int_{-\infty}^{\infty} f_{XY}(x, y) dx dy$$

$$= \int_{-\infty}^{\infty} x f_X(x) dx + \int_{-\infty}^{\infty} y f_Y(y) dy = E[X] + E[Y]$$

Therefore:

[169] $E[X + Y] = E[X] + E[Y]$

The relation in [169] holds if X and Y are statistically independent or not. However, note that $E[XY] \neq E[X]E[Y]$ unless both X and Y are statistically independent. ∎

5.7   Conditional Joint cdf, pdf and Expectation

In this section, we shall expand upon the material developed in section 4.7 to include two random variables. We consider continuous random variables first.

The definition of the conditional joint cdf is:

[170] $F_{X|Y}(X \leq x | Y \leq y) = \dfrac{P(X \leq x \cap Y \leq y)}{P(Y \leq y)}$ &

[171] $F_{Y|X}(Y \leq y | X \leq x) = \dfrac{P(X \leq x \cap Y \leq y)}{P(X \leq x)}$

The conditioning event can also be made to be a range of one of the random variables as follows:

[172] $F_{X|Y}(X \leq x | y_1 < Y \leq y_2) = \dfrac{P(X \leq x \cap y_1 < Y \leq y_2)}{P(y_1 < Y \leq y_2)}$

$$= \dfrac{P(X \leq x \cap y_1 < Y \leq y_2)}{F_Y(y_2) - F_Y(y_1)}$$

The numerator in [172] can be related to the joint cdf using [131] and [129]:

[173] $P(-\infty < X \leq x \cap y_1 < Y \leq y_2) = F_{XY}(x, y_2) + F_{XY}(-\infty, y_1)$

$$-F_{XY}(-\infty, y_2) - F_{XY}(x, y_1) = F_{XY}(x, y_2) + 0 + 0 - F_{XY}(x, y_1)$$

Therefore:

120

[174] $F_{X|Y}(X \leq x | y_1 < Y \leq y_2) = \dfrac{F_{XY}(x, y_2) - F_{XY}(x, y_1)}{F_Y(y_2) - F_Y(y_1)}$

The conditioning event can also be Y=y. However, this could cause a problem if Y is a continuous random variable since, in general, P(Y=y) = 0. This problem is solved by letting Y vary by Δy and then taking the limit as Δy goes to 0 as shown below:

[175] $F_{X|Y}(X \leq x | Y = y) = \lim\limits_{\Delta y \to 0} F_{X|Y}(X \leq x | y < Y \leq y + \Delta y)$

$= \lim\limits_{\Delta y \to 0} \dfrac{P(X \leq x \cap y < Y \leq y + \Delta y)}{P(y < Y \leq y + \Delta y)}$

We can refine the denominator as follows:

$P(y < Y \leq y + \Delta y) = F_Y(y + \Delta y) - F_Y(y)$ Then, using the basic definition of a derivative:

$f_Y(y) = \dfrac{d}{dy} F_Y(y) \cong \dfrac{F_Y(y + \Delta y) - F_Y(y)}{\Delta y}$, which results in

[176] $\Delta y f_Y(y) \cong F_Y(y + \Delta y) - F_Y(y)$

Replacing the denominator of [175] with [176] results in:

[177] $F_{X|Y}(X \leq x | Y = y) = \lim\limits_{\Delta y \to 0} \dfrac{P(X \leq x \cap y < Y \leq y + \Delta y)}{\Delta y f_Y(y)}$

The numerator in [177] can be related to the joint cdf using [131] and [129]:

[178] $P(-\infty < X \leq x, y < Y \leq y + \Delta y) = F_{XY}(x, y + \Delta y) + F_{XY}(-\infty, y)$

$-F_{XY}(-\infty, y + \Delta y) - F_{XY}(x, y) = F_{XY}(x, y + \Delta y) + 0 + 0 - F_{XY}(x, y)$

Note that:

$\dfrac{F_{XY}(x, y + \Delta y) - F_{XY}(x, y)}{\Delta y} \cong \dfrac{\partial}{\partial y} F_{XY}(x, y)$, therefore,

[179] $F_{XY}(x, y + \Delta y) - F_{XY}(x, y) \cong \Delta y \dfrac{\partial}{\partial y} F_{XY}(x, y)$

Inserting [179] into the numerator of [177] results in the conditional cdf:

$$[180] \ F_{X|Y}(X \le x|Y=y) \ = \ \lim_{\Delta y \to 0} \frac{\Delta y \frac{\partial}{\partial y} F_{XY}(x, y)}{\Delta y f_Y(y)} \ = \ \frac{\frac{\partial}{\partial y} F_{XY}(x, y)}{f_Y(y)}$$

Taking the derivative of [180] with respect to x we get the conditional pdf:

$$[181] \ f_{X|Y}(X \le x|Y=y) \ = \ \frac{\partial}{\partial x} F_{X|Y}(X \le x|Y=y) \ = \ \frac{\frac{\partial^2}{\partial x \partial y} F_{XY}(x, y)}{f_Y(y)}$$

$$= \ \frac{f_{XY}(x, y)}{f_Y(y)}$$

For all of the derivations above X and Y can be swapped to obtain the corresponding relations for the other random variable.

Example 5.12:

Given: $f_{XY}(x,y) = kx^2 + y$ where $0<x<1$ and $0<y<1$ find:

a) k
b) the marginal probabilities
c) determine if X and Y are statistically independent
d) $P(0<y\le 1/2|x=1/2)$
a) We use [143] to find k

$$\int_{-\infty}^{\infty} \int_{-\infty}^{\infty} f_{XY}(x, y)dxdy \ = \ 1 \ = \ \int_0^1 \int_0^1 kx^2 + ydxdy$$

$$= \ \int_0^1 k\frac{x^3}{3} + yx \Big|_0^1 dy \ = \ \int_0^1 \frac{k}{3} + ydy \ = \ k\frac{y}{3} + \frac{y^2}{2} \Big|_0^1 \ = \ \frac{k}{3} + \frac{1}{2} \to k \ = \ 1.5$$

b) Next [148] and[149] are used to find the marginal pdfs:

$$f_X(x) = \int_{-\infty}^{\infty} f_{XY}(x, y)dy = \int_0^1 1.5x^2 + ydy = 1.5x^2y + \frac{y^2}{2}\Big|_0^1$$

$$= 1.5x^2 + \frac{1}{2}$$

$$f_Y(y) = \int_{-\infty}^{\infty} f_{XY}(x, y)dx = \int_0^1 1.5x^2 + ydx = 1.5\frac{x^3}{3} + yx\Big|_0^1 = \frac{1}{2} + y$$

c) If X and Y are statistically independent then [157] will hold. Applying [157]:

$$1.5x^2 + y \ne \left(1.5x^2 + \frac{1}{2}\right)\left(\frac{1}{2} + y\right)$$

Since the product of the marginal pdfs does not equal the joint pdf, X and Y are not statistically independent.

d) To solve for the probability, $f_{Y|X}(y|x)$ must be formed by $f_{XY}(x,y)/f_X(x)$; this is [181] with X and Y swapped. Then setting X=1/2 and integrating from Y= 0 to 1/2 the probability can be found as shown below.

$$P\left(0 < Y \le \frac{1}{2}\Big|X=1/2\right) = \int_0^{\frac{1}{2}} \frac{1.5x^2 + y}{1.5x^2 + \frac{1}{2}}\Bigg|_{x = \frac{1}{2}} dy = \int_0^{\frac{1}{2}} \frac{0.375 + y}{0.875}dy$$

$$= \frac{1}{0.875}\left(0.375y + \frac{y^2}{2}\right)\Bigg|_0^{\frac{1}{2}} = 0.375 \quad \blacksquare$$

For discrete random variables we define the conditional cdf of x given $Y=y_k$ as:

$$[182] \quad F_{X|Y}(X=x_i|Y=y_k) = \sum_{i=1}^{N} \frac{P(X=x_i \cap Y=y_k)}{P(Y=y_k)}U(x - x_i)U(y - y_k)$$

The conditional pdf is then:

$$[183] \quad f_{X|Y}(X=x_i|Y=y_k) = \frac{d}{dx}F_{X|Y}(X=x_i|Y=y_k)$$

$$= \sum_{i=1}^{N} \frac{P(X=x_i \cap Y=y_k)}{P(Y=y_k)}\delta(x-x_i)\delta(y-y_k)$$

Example 5.13:

Using the situation from example 5.1 find $f_{X|Y}(X=x_i|Y=2)$. First we restate $F_{XY}(x,y)$: $F_{XY}(x,y)$ = 1/8U(x-0)U(y-0) + 1/4U(x-0)U(y-1) + 1/8U(x-0)U(y-2) + 1/8U(x-1)U(y-0) + 1/4U(x-1)U(y-1) + 1/8U(x-1)U(y-2).

Next we find P(Y=2) = 1/8 + 1/8 = 1/4. i.e. Y=2 in 2 out of 8 possibilities. Now [183] can be used:

$$f_{X|Y}(X=x_i|Y=y_k) = \frac{P(X=0 \cap Y=2)}{P(Y=2)}\delta(x-0)\delta(y-2)$$

$$+ \frac{P(X=1 \cap Y=2)}{P(Y=2)}\delta(x-1)\delta(y-2) = \frac{\frac{1}{8}}{\frac{1}{4}}\delta(x-0)\delta(y-2) + \frac{\frac{1}{8}}{\frac{1}{4}}\delta(x-1)\delta(y-2)$$

$$= \frac{1}{2}\delta(x-0)\delta(y-2) + \frac{1}{2}\delta(x-1)\delta(y-2) \quad \blacksquare$$

Next, we consider conditional mean and variance for discrete random variables. The expected value of Y given X=x_i is determined by the following relationship:

$$[184] \quad E[Y|X=x_i] = \sum_{y_j} y_j \frac{P(Y=y_j \cap X=x_i)}{P(X=x_i)}$$

The conditional variance of Y given X=x_i is given by the relationship below. Going forward we will be adding a subscript to σ to indicate which random variable it is associated with.

$$[185] \quad \sigma^2_{Y|X=x_i} = E\left[(Y - E[Y|X=x_i])^2 \Big| X=x_i\right]$$

$$= \sum_{y_j} (y_j - E[Y|X=x_i])^2 \frac{P(Y = y_j \cap X = x_i)}{P(X = x_i)}$$

[185] can also be expressed as shown in [186] below. This is equivalent to [73] for conditional probabilities.

$$[186] \quad \sigma^2_{Y|X=x_i} = E[Y^2|X=x_i] - [E(Y|X=x_i)]^2$$

Example 5.14:

Considering again the situation from example 5.1, find the conditional expectation and variance of Y given X=1.

Applying [184] we get:

$$E[Y|X=1] = \sum_{y_j} y_j \frac{P(Y = y_j \cap X = x_i)}{P(X = x_i)} = 0 \frac{P(Y = 0 \cap X = 1)}{P(X = 1)}$$

$$+ 1 \frac{P(Y = 1 \cap X = 1)}{P(X = 1)} + 2 \frac{P(Y = 2 \cap X = 1)}{P(X = 1)} = \frac{0\left(\frac{1}{8}\right) + 1\left(\frac{1}{4}\right) + 2\left(\frac{1}{8}\right)}{\frac{1}{2}} = 1$$

To find the conditional variance, [185] is employed:

$$\sigma^2_{Y|X=x_i} = \sum_{y_j} (y_j - E[Y|X=x_i])^2 \frac{P(Y = y_j \cap X = x_i)}{P(X = x_i)}$$

$$= (0-1)^2 \frac{P(Y = 0 \cap X = 1)}{P(X = 1)} + (1-1)^2 \frac{P(Y = 1 \cap X = 1)}{P(X = 1)}$$

$$+ (2-1)^2 \frac{P(Y = 2 \cap X = 1)}{P(X = 1)} = \frac{(-1)^2\frac{1}{8} + (0)^2\frac{1}{4} + (1)^2\frac{1}{8}}{\frac{1}{2}} = \frac{1}{2} \quad \blacksquare$$

For continuous random variables, the relationships for determining the conditional expectation and variance are given below.

[187] $E(Y \leq y | X=x) = \int_{-\infty}^{\infty} y f_{Y|X}(Y \leq y | X=x) dy$

[188] $\sigma^2_{Y|X} = E[(Y - E(Y \leq y | X=x))^2 | x]$

$= \int_{-\infty}^{\infty} (Y - E(Y \leq y | X=x))^2 f_{Y|X}(Y \leq y | X=x) dy$

[189] $\sigma^2_{Y|X} = E[Y^2 | x] - [E(Y|x)]^2$

Example 5.15:

X and Y are distributed as shown in the figure below:

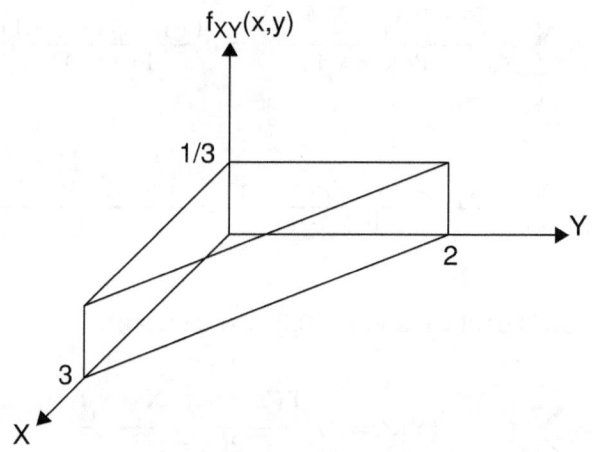

Figure 70 Distribution of X and Y

Find E[X|Y=1]. First we need to determine $f_Y(y)$ using [149]. To do this notice that Y= -(2/3)X +2 which means that for a given Y, X= 3-(3/2)y, thus:

$$f_Y(y) = \int_{-\infty}^{\infty} f_{XY}(x, y)dx = \int_0^{3-\frac{3}{2}y} \frac{1}{3}dx = \frac{1}{3}x\Big|_0^{3-\frac{3}{2}y} = 1 - \frac{1}{2}y$$

Now using [181]:

$$f_{X|Y}(X \le x|Y = y) = \frac{f_{XY}(x, y)}{f_Y(y)} = \frac{\frac{1}{3}}{1 - \frac{1}{2}y}\Bigg|_{y\,=\,1} = \frac{2}{3} \quad 0 < y < 1, \; 0 < x < \frac{3}{2}$$

Next, using [187]:

$$E(X|Y=1) = \int_0^{\frac{3}{2}} x f_{X|Y}(x|y=1)dx = \frac{2}{3}\left(\frac{x^2}{2}\right)\Bigg|_0^{\frac{3}{2}} = \frac{3}{4}$$

Notice that while the integral above gives the correct answer, since $f_{X|Y}(X \le x|Y=1)$ is a uniform distribution, we could have found $E(X|Y=1)$ from inspection of $f_{X|Y}(X \le x|Y=1)$ by using [75] thusly: $E(X|Y=1) = [(3/2) +0] / 2 = 0.75$. That is, when Y=1, X ranges from 0 to 3/2. The average is then (3/2+0)/2=3/4, which agrees with the previous result. ∎

5.8 Total Probability and Baye's Theorem for Two Random Variables

The development of Baye's Theorem for two random variables follows essentially the same process as was introduced in section 2.6.

Let us first look at the continuous case. Restating [181]:

$$[190] \quad f_{X|Y}(X \le x|Y = y) = \frac{f_{XY}(x, y)}{f_Y(y)}$$

Swapping X and Y gives the corresponding conditional pdf as follows:

$$[191] \quad f_{Y|X}(Y \le y|X = x) = \frac{f_{XY}(x, y)}{f_X(x)}$$

Solving for the joint pdfs gives:

[192] $f_{XY}(x, y) = f_{X|Y}(X \leq x | Y = y) f_Y(y)$

[193] $f_{XY}(x, y) = f_{Y|X}(Y \leq y | X = x) f_X(x)$

Next we equate [192] and [193]:

[194] $f_{X|Y}(X \leq x | Y = y) f_Y(y) = f_{Y|X}(Y \leq y | X = x) f_X(x)$

Solving [194] for the conditional pdfs yields the pdf form of Baye's Theorem:

[195] $f_{X|Y}(X \leq x | Y = y) = \dfrac{f_{Y|X}(Y \leq y | X = x) f_X(x)}{f_Y(y)}$

[196] $f_{Y|X}(Y \leq y | X = x) = \dfrac{f_{X|Y}(X \leq x | Y = y) f_Y(y)}{f_X(x)}$

The marginal pdfs in the denominators of [195] and [196] can be obtained through using the joint pdfs in [192] and [193] with [148] and [149] as follows:

[197] $f_X(x) = \displaystyle\int_{-\infty}^{\infty} f_{XY}(x, y) dy = \int_{-\infty}^{\infty} f_{X|Y}(X \leq x | Y = y) f_Y(y) dy$

[198] $f_Y(y) = \displaystyle\int_{-\infty}^{\infty} f_{XY}(x, y) dx = \int_{-\infty}^{\infty} f_{Y|X}(Y \leq y | X = x) f_X(x) dx$

[197] and [198] are used to calculate the pdfs for the total probability. Inserting them into [195] and [196] yields the final forms of Baye's theorem:

[199] $f_{X|Y}(X \leq x | Y = y) = \dfrac{f_{Y|X}(Y \leq y | X = x) f_X(x)}{\displaystyle\int_{-\infty}^{\infty} f_{Y|X}(Y \leq y | X = x) f_X(x) dx}$

[200] $f_{Y|X}(Y \leq y | X = x) = \dfrac{f_{X|Y}(X \leq x | Y = y) f_Y(y)}{\displaystyle\int_{-\infty}^{\infty} f_{X|Y}(X \leq x | Y = y) f_Y(y) dy}$

For the discrete case:

$$[201] \quad P(X{=}x_i | Y{=}y_k) = \frac{P(X{=}x_i \cap Y{=}y_k)}{\sum_i P(X{=}x_i \cap Y{=}y_k)}$$

Note that the marginal probabilities, $P(X{=}x_i)$, $P(Y{=}y_k)$, are found through [153] and [154].

Example 5.16:

Consider the following from example 5.1:

$P(X{=}x_0 \cap Y{=}y_0) = 1/8$

$P(X{=}x_0 \cap Y{=}y_1) = 1/4$

$P(X{=}x_0 \cap Y{=}y_2) = 1/8$

$P(X{=}x_1 \cap Y{=}y_0) = 1/8$

$P(X{=}x_1 \cap Y{=}y_1) = 1/4$

$P(X{=}x_1 \cap Y{=}y_2) = 1/8$

Find $P(X{=}x_1 | Y{=}y_2)$. To do this we apply [154] and [201]:

$$P(Y = y_2) = \sum_{i=0}^{1} P(X = x_i \cap Y = y_2) = \frac{1}{8} + \frac{1}{8} = \frac{1}{4}$$

$$P(X{=}x_1 | Y{=}y_2) = \frac{P(X{=}x_1 \cap Y{=}y_2)}{P(Y{=}y_2)} = \frac{\frac{1}{8}}{\frac{1}{4}} = \frac{1}{2} \quad \blacksquare$$

5.9   Correlation and Covariance

The expected value of the random variables X and Y is called the correlation of random variables X and Y and is shown below:

[202] $E[XY] = \displaystyle\int_{-\infty}^{\infty} \int_{-\infty}^{\infty} xyf_{XY}(x, y)dxdy$, continuous case

[203] $E[XY] = \displaystyle\sum_{x_i} \sum_{y_j} x_i y_j P(X = x_i \cap Y = y_j)$, discrete case

If E[XY]=0 we say that X and Y are orthogonal.

The covariance of X and Y is defined as:

[204] $\sigma_{XY} = E[(X - \bar{X})(Y - \bar{Y})] = E[XY - \bar{X}Y - X\bar{Y} + \bar{X}\bar{Y}]$

$= E[XY] - E[\bar{X}Y] - E[X\bar{Y}] + E[\bar{X}\bar{Y}] = E[XY] - \bar{X}E[Y] - \bar{Y}E[X] + E[\bar{X}\bar{Y}]$

$= E[XY] - E[X]E[Y] - E[Y]E[X] + E[X]E[Y] = E[XY] - E[X]E[Y]$

If σ_{XY} is equal to 0, we say that X and Y are uncorrelated. That is:

[205] $E[XY] = E[X]E[Y]$

If X and Y are independent then they are also uncorrelated.

Example 5.17:

Prove that if X and Y are independent, then they are also uncorrelated.

Using [157] and [164] we can write:

$$E[XY] = \int_{-\infty}^{\infty} \int_{-\infty}^{\infty} xyf_{XY}(x, y)dxdy = \int_{-\infty}^{\infty} \int_{-\infty}^{\infty} xyf_X(x)f_Y(y)dxdy$$

We evaluate the integral over the variable x first. In this case y and $f_Y(y)$ are constants and can be put outside of the integral as shown below:

$$E[XY] = \int_{-\infty}^{\infty} yf_Y(y) \int_{-\infty}^{\infty} xf_X(x)dxdy$$

Note that $\displaystyle\int_{-\infty}^{\infty} x f_X(x) dx$ is a constant with respect to the y variable.

So when integrating with respect to y, this integral can be pulled outside of the left most integral as shown below and results in [205]:

$$E[XY] = \int_{-\infty}^{\infty} x f_X(x) dx \int_{-\infty}^{\infty} y f_Y(y) dy = E[X]E[Y]$$

Note, however, that while independent random variables are uncorrelated, this does not mean that uncorrelated random variables are necessarily independent.

Just for completeness, we will also look at the discrete case. Using [168] and the result of [158] we can write:

$$E[XY] = \sum_{x_i}\sum_{y_j} x_i y_j P(X = x_i \cap Y = y_j)$$

$$= \sum_{x_i}\sum_{y_j} x_i y_j P(X = x_i) P(Y = y_j)$$

Note that for each given x_i, a summation is done over each y_j. This allows the above summation to be written as:

$$E[XY] = \sum_{x_i} x_i P(X = x_i) \sum_{y_j} y_j P(Y = y_j) = E[X]E[Y]$$

Remember, this result applies to independent random variables. ∎

5.10 pdf of a Function of Two Random Variables

Here we consider how to determine the pdf of a function of two random variables. The 1st case that we consider is the sum of two random variables X and Y. We define this as Z=X+Y. Based on the material in section 4.3, the cdf of Z is simply:

[206] $F_Z(z) = P(Z \le z) = P(X + Y \le z)$

To obtain $F_Z(z)$ we integrate $f_{XY}(x,y)$ as in [144]. However, we need to select the integration limits correctly. Referring to Figure 71 below, the limit of integration for x can be seen to be from -∞ to z-y to cover the range of x, while for y, the limits are -∞ to +∞ in order to cover the full range of y. Thus, we end up with:

[207] $F_Z(z) = \int_{-\infty}^{\infty} \int_{-\infty}^{z-y} f_{XY}(x, y) dx dy$

The pdf can then be found through differentiating [207] with respect to z:

[208] $f_Z(z) = \frac{d}{dz} F_Z(z)$

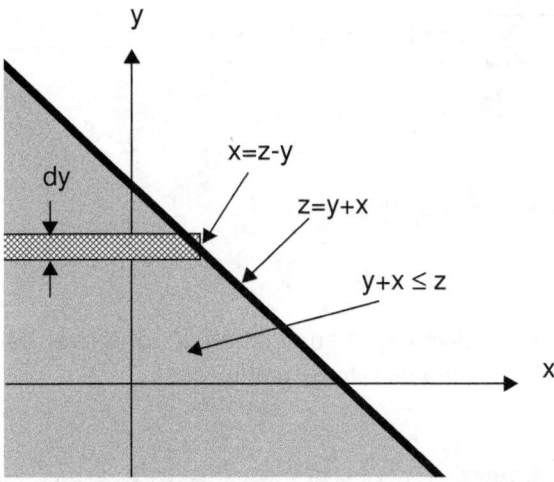

Figure 71 Determining the Limits of Integration

Now if X and Y are statistically independent, then [207] becomes:

[209] $F_Z(z) = \int_{-\infty}^{\infty} \int_{-\infty}^{z-y} f_X(x) f_Y(y) dx dy$

When performing the integration with respect to x, $f_Y(y)$ is a constant, therefore, it can be brought outside of the inner integral as shown:

$$[210] \quad F_Z(z) = \int_{-\infty}^{\infty} f_Y(y) \int_{-\infty}^{z-y} f_X(x)\, dx\, dy$$

The pdf can then be found by differentiating [210] with respect to z. This can be done using Leibnitz's rule[1]. In our case z will replace λ, $v(z) = \infty$, $u(z) = -\infty$, and $f(y, z) = f_Y(y) \int_{-\infty}^{z-y} f_X(x)\, dx$. Let's take each of the terms individually.

$$[211] \quad f(v(z), z)\frac{\partial}{\partial z}v(z) = \left(f_Y(\infty) \int_{-\infty}^{z - \infty} f_X(x)\, dx \right)\left(\frac{\partial \infty}{\partial z} \right) = 0$$

This is because each of the terms in [211] are 0. $f_Y(\infty) = 0$, because $F_Y(\infty) = 1$ and $f_Y(\infty)$ is thus the derivative of 1 which is 0. The integral is 0 because no matter what z is, the limits of the integral are always going to be $-\infty$. The partial derivative is also zero because ∞ can be considered to be a constant. Next we look at:

$$[212] \quad f(u(z), z)\frac{\partial}{\partial z}v(z) = \left(f_Y(-\infty) \int_{-\infty}^{(z + \infty)} f_X(x)\, dx \right)\left(\frac{\partial}{\partial z}(-\infty) \right) = 0$$

Let us now examine the terms of [212]. $f_Y(-\infty) = 0$, because $F_Y(-\infty) = 0$ and $f_Y(-\infty)$ is thus the derivative of 0 which is 0. The integral is 1 because no matter what z is, the limits of the integral are always going to be $-\infty$ to $+\infty$ and the integral of the marginal pdf from $-\infty$ to $+\infty$ is 1. The partial derivative is also zero because $-\infty$ can be considered to be a constant. So 0*1*0=0. Next we consider the final term:

$$[213] \quad \int_{u(z)}^{v(z)} \left(\frac{\partial}{\partial z}f(y, z) \right) dy = \int_{-\infty}^{\infty} f_Y(y)\frac{\partial}{\partial z}\left[\int_{-\infty}^{z-y} f_X(x)\, dx \right] dy$$

---

1. $\dfrac{\partial}{\partial \lambda}\left[\displaystyle\int_{u(\lambda)}^{v(\lambda)} f(y, \lambda)\, dy \right] = \displaystyle\int_{u(\lambda)}^{v(\lambda)} \left(\frac{\partial}{\partial \lambda}f(y, \lambda) \right) dy + f(v(\lambda), \lambda)\frac{\partial}{\partial \lambda}v(\lambda)$

$-f(u(\lambda), \lambda)\dfrac{\partial}{\partial \lambda}u(\lambda)$.

Note that $f_Y(y)$ can be brought outside the partial derivative since it is considered a constant as far as the partial derivative is concerned. To finish evaluating [213] we need use Leibnitz's rule one more time on the bracketed term. Thus,

$$[214] \quad \frac{\partial}{\partial z}\left[\int_{-\infty}^{z-y} f_X(x)dx\right] = \int_{-\infty}^{z-y}\left[\frac{\partial}{\partial z}f_X(x)\right]dx + f_X(z-y)\frac{\partial}{\partial z}(z-y)$$

$$- f_X(-\infty)\frac{\partial}{\partial z}(-\infty) = 0 + f_X(z-y) - 0 = f_X(z-y)$$

Inserting the result of [214] into [213] results in:

$$[215] \quad \int_{u(z)}^{v(z)}\left(\frac{\partial}{\partial z}f(y,z)\right)dy = \int_{-\infty}^{\infty} f_Y(y)f_X(z-y)dy$$

Summing the results of [211], [212] and [215] results in the derivative of [210]:

$$[216] \quad f_Z(z) = \int_{-\infty}^{\infty} f_Y(y)f_X(z-y)dy$$

This is readily recognized as a convolution integral. Therefore, we can conclude that the density function of the sum of two independent random variables is the convolution of their marginal pdfs. Don't forget that this conclusion applies to independent random variables. Additionally, note that we can expand this to cases where more than two independent random variables are summed. Consider $Y_1 = X_1 + X_2$ so that $f_{Y1}(y) = X_1 * X_2$ Next consider a third independent random variable X_3. X_3 is independent of Y_1 because it is independent of X_1 and X_2. So we can convolve Y_1 and X_3 to create $f_{Y2}(y)$ as follows $f_{Y2}(y) = Y_1 * X_3 = X_1 * X_2 * X_3$, where * represents the convolution operation. Repeating this line of thought, n independent random variables can be handled: $f_Y(y) = X_1 * X_2 * \ldots X_n$.

Example 5.18:

To demonstrate how to apply [216] let us consider two independent pdfs:

$$f_X(x) = 2 \ \ 0 \le x \le \frac{1}{2} \ \text{ and } f_Y(y) = y \ \ 0 \le y \le \sqrt{2}$$

Inserting these pdfs into [216] is straight forward. The part that needs careful consideration is determining the limits of integration. To do this, consider the figures below[1]. $f_Y(y)$ is plotted along with $f_X(z-y)$, which is just a mirror image of $f_X(y)$ shifted by z. The range of z is determined by Z=X+Y. Zmin= Xmin+Ymin= 0 and Zmax = Xmax + Ymax = $1/2 + 2^{1/2}$.

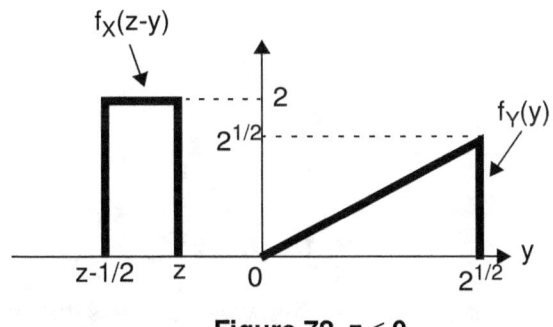

Figure 72 z ≤ 0

In Figure 72 above, it can be seen that when z ≤ 0 the product of the two pdfs is zero. So $f_Z(z)$ is 0 for z ≤ 0.

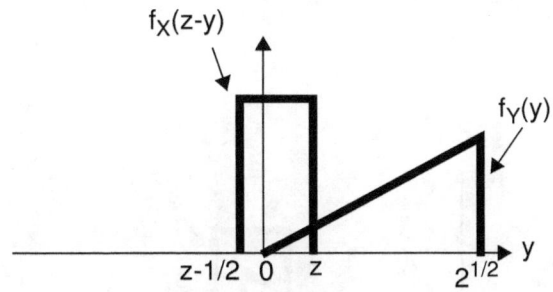

Figure 73 0 < z ≤ 1/2

In Figure 73 above, it can be seen that when z is between 0 and 1/2 the limit of integration can be seen to be from 0 to z as follows:

---

1. The shapes are not drawn to scale. They are intended to show basic relationships.

$$f_Z(z) = \int_0^z 2y\,dy = z^2 \quad 0 < z \le \frac{1}{2}$$

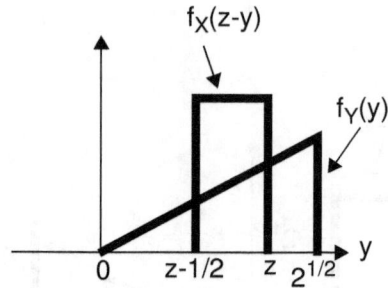

Figure 74 $1/2 \le z \le 2^{1/2}$

In Figure 74 above, it can be seen that when z is less than or equal to $2^{1/2}$ but greater or equal to than 1/2 the product of the two pdfs is non-zero. The limit of integration can thus be seen to be from z-1/2 to z as follows:

$$f_Z(z) = \int_{z-\frac{1}{2}}^{z} 2y\,dy = z - \frac{1}{4} \quad 1/2 \le z \le \sqrt{2}$$

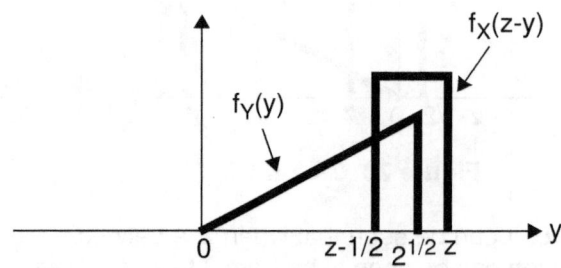

Figure 75 $2^{1/2} \le z \le 1/2 + 2^{1/2}$

In Figure 75 above, it can be seen that when z is less than or equal to $1/2+2^{1/2}$ but greater than or equal to $2^{1/2}$ the limit of integration can be seen to be from z-1/2 to $2^{1/2}$ as follows:

$$f_Z(z) = \int_{z-\frac{1}{2}}^{\sqrt{2}} 2y\,dy = -z^2 + z + \frac{7}{4} \quad \sqrt{2} \le z \le \frac{1}{2} + \sqrt{2}$$

Finally, it should be clear that for z greater than $2+2^{1/2}$ the two pdfs no longer overlap and $f_Z(z)$ is 0.

Summarizing the results:

$$f_Z(z) = \begin{cases} 0 & z \le 0 \\ z^2 & 0 \le z \le \frac{1}{2} \\ z - \frac{1}{4} & \frac{1}{2} \le z \le \sqrt{2} \\ -z^2 + z + \frac{7}{4} & \sqrt{2} < z \le \frac{1}{2} + \sqrt{2} \\ 0 & z > \frac{1}{2} + \sqrt{2} \end{cases}$$

This example may seem to have been tedious, but it shows that due to the convolution operation of sliding $f_X(z-y)$ over $f_Y(y)$, care needs to be taken to select the correct limits of integration. As a very important point, we can check our answer by calculating the mean of Z using $f_Z(z)$ and the mean of X+Y using $f_{XY}(x,y)$; both results should match. The equations for both methods are set up below, showing the results:

$$E[Z] = \int_0^{\frac{1}{2}+\sqrt{2}} z f_Z(z)\,dz = \int_0^{\frac{1}{2}} z^3\,dz + \int_{\frac{1}{2}}^{\sqrt{2}} \left(z^2 - \frac{z}{4}\right)dz$$

$$+ \int_{\sqrt{2}}^{\frac{1}{2}+\sqrt{2}} \left(-z^3 + z^2 + \frac{7}{4}z\right)dz = \left(\frac{1}{64}\right) + \left(\frac{2\sqrt{2}}{3} - \frac{25}{96}\right) + \left(\frac{95}{192}\right) = \frac{2\sqrt{2}}{3} + \frac{1}{4}$$

$$E[X + Y] = \int_0^{\sqrt{2}} \int_0^{\frac{1}{2}} (x + y) f_{XY}(x, y) dxdy$$

$$= \int_0^{\sqrt{2}} \int_0^{\frac{1}{2}} 2y(x + y) dxdy = \frac{2\sqrt{2}}{3} + \frac{1}{4}$$

Since E[Z] = E[X+Y] the calculation of $f_Z(z)$ checks out[1]. Again, remember that the application of [216] can only be made if X and Y are statistically independent. Otherwise, [207] and [208] must be used instead. ∎

As a final topic in this section, we shall now consider two functions of two random variables. Let's start with two random variables, X and Y for which we know their joint pdf, $f_{XY}(x,y)$. We can define two additional random variables as functions of X and Y; let's call them Z=g(X,Y) and W=h(X,Y). We also define inverse relations, X=k(W,Z) and Y=m(W,Z).

Our object now is to determine the joint pdf of W and Z which we will call $f_{WZ}(w,z)$. We take advantage of the fact that the probability that P($w_1 \leq W \leq w_2$ ∩ $z_1 \leq Z \leq z_2$) must equal P($x_1 \leq X \leq x_2$ ∩ $y_1 \leq Y \leq y_2$). Thus, using [147]:

$$[217] \quad \int_{y_1}^{y_2} \int_{x_1}^{x_2} f_{XY}(x, y) dxdy = \int_{z_1}^{z_2} \int_{w_1}^{w_2} f_{WZ}(w, z) dwdz$$

We can use the change of variables formula[2] to then write:

---

1. Note that [169] could have been used to calculate E[X+Y].

2. $\displaystyle\iint_R f(x, y) dxdy = \iint_{R*} f(x(u, v), y(u, v)) \left| \frac{\partial(x, y)}{\partial(u, v)} \right| dudv$ where   x=x(u,v),

y=y(u,v) and $\dfrac{\partial(x, y)}{\partial(u, v)} = \begin{vmatrix} \dfrac{\partial x}{\partial u} & \dfrac{\partial x}{\partial v} \\ \dfrac{\partial y}{\partial u} & \dfrac{\partial y}{\partial v} \end{vmatrix}$. The matrix part is called the Jacobian.

[218] $\displaystyle\int_{y_1}^{y_2}\int_{x_1}^{x_2} f_{XY}(x, y)dxdy$

$\displaystyle= \int_{z_1}^{z_2}\int_{w_1}^{w_2} f_{XY}(k(w, z), m(w, z))\left|\frac{\partial(x, y)}{\partial(w, z)}\right| dwdz$

By visual inspection of [217] and [218] we can see that:

[219] $\displaystyle f_{WZ}(w, z) = f_{XY}(k(w, z), m(w, z))\left|\frac{\partial(x, y)}{\partial(w, z)}\right|$

$\displaystyle= f_{XY}(k(w, z), m(w, z))\left|\begin{bmatrix}\dfrac{\partial x}{\partial w} & \dfrac{\partial x}{\partial z}\\[2mm] \dfrac{\partial y}{\partial w} & \dfrac{\partial y}{\partial z}\end{bmatrix}\right|$

Therefore if we have two random variables and their joint pdf, we can calculate the joint pdf of two functions of those random variables.

Example 5.19:

We are given two random variables X and Y and their joint pdf $f_{XY}(x,y)$. There are two random variables defined as W=X (for simplicity) and Z=X+Y. We wish to find the joint pdf of W and Z. This is done by applying [219].

X = k(W,Z) = W

Y = m(W,Z) = Z - X = Z - W

Next we evaluate the matrix part $\left|\begin{bmatrix}\dfrac{\partial x}{\partial w} & \dfrac{\partial x}{\partial z}\\[2mm] \dfrac{\partial y}{\partial w} & \dfrac{\partial y}{\partial z}\end{bmatrix}\right| = \begin{bmatrix}1 & 0\\ -1 & 1\end{bmatrix} = 1 - 0 = 1$

Inserting X, Y, and the absolute value of the matrix into [219] results in:

[220] $\displaystyle f_{WZ}(w, z) = f_{XY}(k(w, z), m(w, z))\left|\frac{\partial(x, y)}{\partial(w, z)}\right| = f_{XY}(w, z - w)$

Keep in mind that if desired, the marginal pdf of z and w can also be found using [148] as shown below:

$$f_Z(z) = \int_{-\infty}^{\infty} f_{WZ}(w, z)dw \quad \text{and} \quad f_W(w) = \int_{-\infty}^{\infty} f_{WZ}(w, z)dz.$$ How-

ever, note that in order to determine the correct limits of integration, the area over which $f_{WZ}(w,y)$ exists must be drawn. See Exercise 9. ■

5.11   Joint Characteristic Function

The joint characteristic function is defined as:

[221] $$\Phi_{XY}(x, y) = E\left[e^{j\omega_1 X + j\omega_2 Y}\right] = E\left[e^{j\omega_1 X} e^{j\omega_2 Y}\right]$$

where of course, $j = \sqrt{-1}$.

Thus, for the continuous case,

[222] $$\Phi_{XY}(\omega_1, \omega_2) = E\left[e^{j\omega_1 X} e^{j\omega_2 Y}\right]$$

$$= \int_{-\infty}^{\infty} \int_{-\infty}^{\infty} e^{j\omega_1 x} e^{j\omega_2 y} f_{XY}(x, y)dxdy$$

Note that the marginal characteristic functions are obtained by:

[223] $$\Phi_X(\omega_1) = \Phi_{XY}(\omega_1, 0)$$

[224] $$\Phi_Y(\omega_2) = \Phi_{XY}(0, \omega_2)$$

As was done in section 4.9, we can determine joint moments of X and Y through operating on the joint characteristic function:

[225] $$E[X^n Y^m] = \overline{X^n Y^m} = \frac{1}{j^{n+m}} \frac{\partial^{(n+m)}}{\partial\omega_1^n \partial\omega_2^m} \Phi_{XY}(\omega_1, \omega_2)$$

To arrive at the relationship in [225], we first use Leibnitz's rule[1] to differentiate [222] with respect to ω_1 which results in:

[226] $$\frac{\partial^1}{\partial\omega_1^1 \partial\omega_2^0}\Phi_{XY}(\omega_1, \omega_2) = \int_{-\infty}^{\infty} \frac{\partial}{\partial\omega_1}\int_{-\infty}^{\infty} e^{j\omega_1 x} e^{j\omega_2 y} f_{XY}(x, y)dxdy$$

Applying Leibnitz's rule again to [226] to finish the job:

[227] $$\frac{\partial^1}{\partial\omega_1^1 \partial\omega_2^0}\Phi_{XY}(\omega_1, \omega_2) = \int_{-\infty}^{\infty}\int_{-\infty}^{\infty} \frac{\partial}{\partial\omega_1} e^{j\omega_1 x} e^{j\omega_2 y} f_{XY}(x, y)dxdy$$

$$= \int_{-\infty}^{\infty}\int_{-\infty}^{\infty} jx e^{j\omega_1 x} e^{j\omega_2 y} f_{XY}(x, y)dxdy$$

Repeatedly applying Leibnitz's rule to differentiate [227] n times with respect to ω_1 results in:

[228] $$\frac{\partial^n}{\partial\omega_1^n \partial\omega_2^0}\Phi_{XY}(\omega_1, \omega_2) = \int_{-\infty}^{\infty}\int_{-\infty}^{\infty} j^n x^n e^{j\omega_1 x} e^{j\omega_2 y} f_{XY}(x, y)dxdy$$

Using Leibnitz's rule to differentiate [228] m times with respect to ω_2 results in:

[229] $$\frac{\partial^{n+m}}{\partial\omega_1^n \partial\omega_2^m}\Phi_{XY}(\omega_1, \omega_2)$$

$$= \int_{-\infty}^{\infty}\int_{-\infty}^{\infty} (j^{n+m})x^n y^m e^{j\omega_1 x} e^{j\omega_2 y} f_{XY}(x, y)dxdy$$

Dividing [229] by j^{n+m} and setting both ω_1 and ω_2 to 0 results in:

---

1. $\frac{\partial}{\partial\lambda}\int_a^b f(x, \lambda)dx = \int_a^b \frac{\partial}{\partial\lambda}f(x, \lambda)dx$.

[230] $\dfrac{1}{j^{n+m}}\dfrac{\partial^{n+m}}{\partial\omega_1^n\partial\omega_2^m}\Phi_{XY}(\omega_1,\omega_2)\Bigg|_{\omega_1=\omega_2=0}$

$$= \int_{-\infty}^{\infty}\int_{-\infty}^{\infty} x^n y^m f_{XY}(x,y)\,dx\,dy$$

The right hand side of [230] is recognized as the expected value of $X^n Y^m$, $E[X^n Y^m]$, which proves [225].

Example 5.20:

Let us consider two random variables X and Y that have a jointly uniform pdf of 1/4 from $0 \le x,y \le 2$. We will calculate the expected value of $X^2 Y$ using [164] and then [225] so that we can verify that we get the same results.

Using [164]:

[231] $E[X^2Y] = \displaystyle\int_0^2\int_0^2 \dfrac{x^2 y}{4}\,dx\,dy = \dfrac{1}{4}\left[\dfrac{x^3}{3}\Big|_0^2\right]\left[\dfrac{y^2}{2}\Big|_0^2\right] = \left(\dfrac{1}{4}\right)\left(\dfrac{8}{3}\right)\left(\dfrac{4}{2}\right) = \dfrac{4}{3}$

Using [225]:

First we need to calculate the joint characteristic function using [221],

$$\Phi_{XY}(\omega_1,\omega_2) = \int_0^2\int_0^2 \dfrac{e^{j\omega_1 x}e^{j\omega_2 y}}{4}\,dx\,dy = \dfrac{1}{4}\left[\dfrac{e^{j\omega_1 x}}{j\omega_1}\Big|_0^2\right]\left[\dfrac{e^{j\omega_2 y}}{j\omega_2}\Big|_0^2\right]$$

$$= \dfrac{1}{4}\left[\dfrac{e^{j\omega_1 2}-1}{j\omega_1}\right]\left[\dfrac{e^{j\omega_2 2}-1}{j\omega_2}\right]$$

Now that the characteristic function has been calculated, [225] can be used to find the desired expected value[1]:

---

1. Due to the complexity of the differentiation and the likelihood of making a mistake, a symbolic math program was used to evaluate the derivative.

[232] $\dfrac{1}{j^3}\dfrac{\partial^3}{\partial\omega_1^2\partial\omega_2^1}\Phi_{XY}(\omega_1,\omega_2)$

$$= -\left[\dfrac{je^{2j\omega_2}}{2\omega_2}+\dfrac{1-e^{2j\omega_2}}{4\omega_2^2}\right]\left[\dfrac{2j\left(e^{2j\omega_1}-1\right)}{\omega_1^3}+\dfrac{4e^{2j\omega_1}}{\omega_1^2}-\dfrac{4je^{2j\omega_1}}{\omega_1}\right]$$

Next setting both ω_1 and ω_2 to 0[1] results in [232] becoming equal to 4/3 as was calculated in [231]. ∎

To wrap up this section, we present the joint characteristic function for the discrete case:

[233] $\Phi_{XY}(\omega_1,\omega_2) = E\left[e^{j\omega_1 X + j\omega_2 Y}\right]$

$$= \sum_{x_i}\sum_{y_j} e^{j\omega_1 x_i + j\omega_2 y_j} P(X = x_i \cap Y = y_j)$$

5.12 Exercises

1 There are two random variables X and Y. The pmf of X is:

$$f_X(x) = 0.3\delta(x) + 0.3\delta(x-1) + 0.4\delta(x-2)$$

The joint pmf of X and Y is given by the figure below:

---

1. A symbolic math program was also used to evaluate [232] for ω_1 and ω_2 set to 0. L'Hôpital's rule could have also been used to evaluate the indeterminate forms in [232]. See reference [10], page 119, equation 4.7-9.

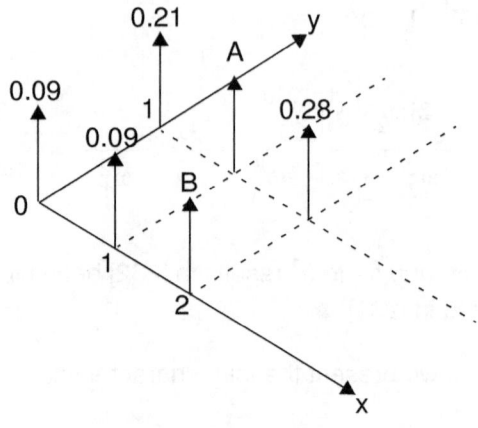

Figure 76 Joint pdf for Exercise 1

What is the pmf of Y?

2 Is the following a valid cdf?

$$F_{XY}(x, y) = (1 - e^{-x})(1 - e^{-y}), \ 0 \le x < \infty, \ 0 \le y < \infty$$

3 Consider the uniformly distributed pdf in the figure below:

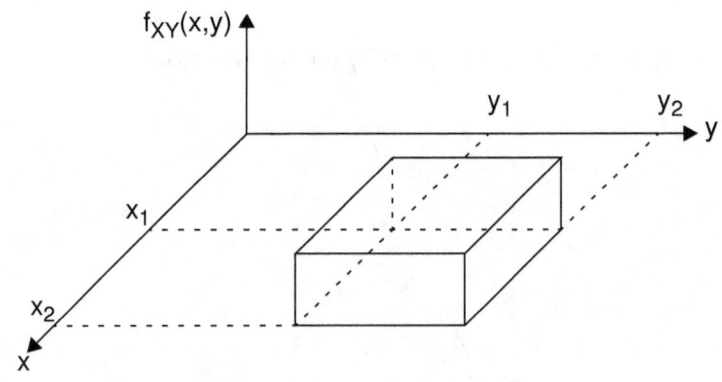

Figure 77 pdf for Exercise 3

Find the cdf.

4 Given $f_{XY}(x,y) = kx^2y$ where $0<x<1$ and $0<y<1$, determine if X and Y are statistically independent.

5 Find $E[Y^2|X=1]$ for example 5.1.

6 An experiment on two random variables X and Y shows that they are uniformly distributed over the area, R, shown in the figure below.

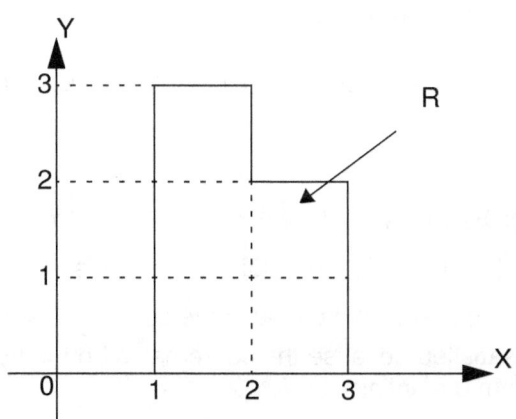

Figure 78

a) Find $f_{XY}(x,y)$.
b) Find $E[X|Y=1]$.
c) Find $E[X|Y=2.5]$.
d) Find $E[Y|X=1]$.

7 $f_{XY}(x,y) = (15/32)(x^2 + y^4)$ where x and y range from -1 to 1. Show that X and Y are uncorrelated but not independent.

8 Given two independent random variables, X and Y that are uniformly distributed from 2 to 3 and from 1 to 4 respectively, find the pdf of their sum.

9 Given two random variables, X and Y that are uniformly distributed from 0 to 1. Two new random variables are defined Z= X + Y and W= X:

a) Find $f_{WZ}(w,z)$.
b) Find the marginal pdf for z (i.e $f_Z(z)$). Hint: Draw the area over which $f_{WZ}(w,z)$ exists to determine the correct limits of integration.

10 For the situation in example 5.1, find the joint characteristic function.

5.13   Solutions to Exercises

1 First lets find A and B:

$$f_X(x) = 0.3\delta(x) + 0.3\delta(x-1) + 0.4\delta(x-2) = 0.3\delta(x) + (0.09 + A)\delta(x-1)$$

$$+ (B + 0.28)\delta(x-2) \quad A \to 0.21 \ \& \ B \to 0.12$$

Now the pmf for Y can be found by:

$$f_Y(y) = (0.09 + 0.09 + 0.12)\delta(y) + (0.21 + 0.21 + 0.28)\delta(y-1)$$

$$= 0.3\delta(y) + 0.7\delta(y-1)$$

2 We are given: $F_{XY}(x, y) = (1 - e^{-x})(1 - e^{-y})$, $0 \le x < \infty$, $0 \le y < \infty$. To be a valid cdf, [127]-[131], [134], and [135] must be satisfied.

- [127] - [129] is satisfied because the cdf is 0 for x or y less than 0.
- [130] is satisfied because the cdf remains bound from 0 to 1 as x and y vary from 0 to infinity.
- [134] is satisfied because the cdf increases from 0 to 1 as x and y increase from 0 to infinity.
- [135] is satisfied because at x and y equal to infinity the cdf is equal to 1.
- [131] is also satisfied. Let's set x_1=3, x_2=25, y_1=4 and y_2=15.

$$P(x_1 < X \le x_2 \cap y_1 < Y \le y_2) = F_{XY}(x_2, y_2) + F_{XY}(x_1, y_1)$$

$$-F_{XY}(x_1, y_2) - F_{XY}(x_2, y_1) = 9.119(10^{-4})$$

Now as a check, we use [147]:

$$P(3 < X \le 25 \cap 4 < Y \le 15) = \int_4^{15} \int_3^{25} e^{-x}e^{-y}dxdy = 9.119(10^{-4})$$

Since the answer matches, we can conclude that the given cdf is valid since it meets [127]-[131], [134], and [135].

3 If the joint pdf is known, then [144] can be used to find the cdf. Since the pdf is uniformly distributed:

146

$$f_{XY}(x, y) = \frac{1}{(x_2 - x_1)(y_2 - y_1)}, \quad x_1 \le x \le x_2 \ \& \ y_1 \le y \le y_2$$

Applying [144]:

$$F_{XY}(x, y) = \int_{y1}^{y} \int_{x1}^{x} \frac{1}{(x_2 - x_1)(y_2 - y_1)} d\xi_1 d\xi_2 = \frac{\left(\xi_1\Big|_{x_1}^{x}\right)\left(\xi_2\Big|_{y_1}^{y}\right)}{(x_2 - x_1)(y_2 - y_1)}$$

$$= \frac{(x - x_1)(y - y_1)}{(x_2 - x_1)(y_2 - y_1)}$$

Now as a check, we can apply [139] to see if we get the original pdf.

$$\frac{\partial}{\partial x} \frac{(x - x_1)(y - y_1)}{(x_2 - x_1)(y_2 - y_1)} = \frac{(y - y_1)}{(x_2 - x_1)(y_2 - y_1)}$$

then

$$\frac{\partial}{\partial y} \frac{(y - y_1)}{(x_2 - x_1)(y_2 - y_1)} = \frac{1}{(x_2 - x_1)(y_2 - y_1)}$$

Which is the original pdf as expected.

4 $f_{XY}(x,y) = kx^2 y$ where $0 < x < 1$ and $0 < y < 1$. First we need to find k using [143].

$$\int_0^1 \int_0^1 kx^2 y \, dx \, dy = 1 = k\left(\frac{x^3}{3}\Big|_0^1\right)\left(\frac{y^2}{2}\Big|_0^1\right) = k\left(\frac{1}{3}\right)\left(\frac{1}{2}\right) \rightarrow k = 6$$

$$\therefore f_{XY}(x, y) = 6x^2 y.$$

The next step is to find the marginal pdfs using [148] and [149]:

$$f_X(x) = \int_0^1 6x^2 y \, dy = 6x^2 \left(\frac{y^2}{2} \bigg|_0^1 \right) = 3x^2$$

$$f_Y(y) = \int_0^1 6x^2 y \, dx = 6y \left(\frac{x^3}{3} \bigg|_0^1 \right) = 2y$$

Since $f_{XY}(x, y) = 6x^2 y = f_X(x)f_Y(y)$, we can conclude that X and Y are independent.

5

This problem can be solved two ways. The 1st step is to adapt [184] to incorporate y_j^2:

$$E[Y^2|X{=}1] = \sum_{y_j} y_j^2 \frac{P(Y = y_j \cap X = x_i)}{P(X = x_i)} = 0^2 \frac{P(Y = 0 \cap X = 1)}{P(X = 1)}$$

$$+ 1^2 \frac{P(Y = 1 \cap X = 1)}{P(X = 1)} + 2^2 \frac{P(Y = 2 \cap X = 1)}{P(X = 1)} = \frac{0\left(\frac{1}{8}\right) + 1\left(\frac{1}{4}\right) + 4\left(\frac{1}{8}\right)}{\frac{1}{2}} = \frac{3}{2}$$

The 2nd way is to insert the values from example 5.14 into [186]:

$$\sigma^2{}_{Y|X=1} = E[Y^2|X{=}1] - [E(Y|X{=}1)]^2 = \frac{1}{2} = E[Y^2|X{=}1] - 1^2$$

Solving for $E[Y^2|X{=}1]$ we end up with 3/2 as above.

6

a) To find $f_{XY}(x,y)$ we note that [143] tells us that the volume of a joint pdf is 1. Therefore, since the area covered by X and Y is 5, and the pdf is uniform, the magnitude of the pdf must be 1/5. Therefore:

$$f_{XY}(x, y) = \frac{1}{5} \text{ for } x,y \in R$$

b) To find E[X|Y=1], we need to find $f_{XY}(X \le x|Y=1)$ using [181]. Even before that, we need to find $f_Y(y)$ using [149].

$$f_Y(y) = \int_1^3 \frac{1}{5}dx = \frac{2}{5}$$

$$f_{X|Y}(X \le x|Y= 1) = \frac{f_{XY}(x, y)}{f_Y(y)} = \frac{\frac{1}{5}}{\frac{2}{5}} = \frac{1}{2} \text{ Then adapting [187]:}$$

$$E(X \le x|Y=1) = \int_1^3 x f_{X|Y}(X \le x|Y=1)dx = \frac{1}{2}\left(\frac{x^2}{2}\Big|_1^3\right) = \frac{1}{2}\left(\frac{9}{2}-\frac{1}{2}\right) = 2$$

Let's solve this problem in a different way. If we were to look into the X axis at the cross section formed by Y=1 we would see the following:

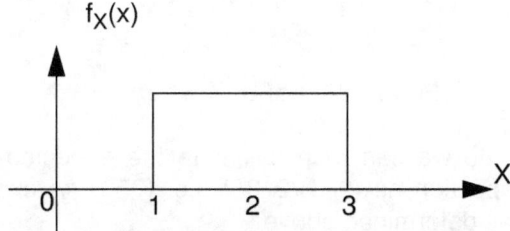

Figure 79 pdf of X when Y=1

From Figure 79 we can then calculate the expected value of X when Y=1 by inspection as (3+1)/2=2 (i.e. [75]). This is the same value as determined above, but as you can see, this way was much more intuitive.

c) To find E[X|Y=2.5] we use the two methods as shown above. Using [149].

$$f_Y(y) = \int_1^2 \frac{1}{5}dx = \frac{1}{5}$$

$$f_{X|Y}(X \le x|Y = 2.5) = \frac{f_{XY}(x, y)}{f_Y(y)} = \frac{\frac{1}{5}}{\frac{1}{5}} = 1 \text{, then adapting [187]:}$$

$$E(X \le x|Y=2.5) = \int_1^2 xf_{X|Y}(X \le x|Y= 2.5)dx = 1\left(\frac{x^2}{2}\Big|_1^2\right) = 1\left(\frac{4}{2}-\frac{1}{2}\right) = \frac{3}{2}$$

As a check, looking into the X axis at the cross section formed by Y=2.5 we would see the following:

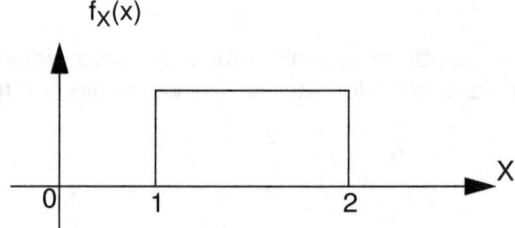

Figure 80 pdf of X when Y=2.5

From Figure 80 we can then calculate the expected value of X when Y=2.5 by inspection as (2+1)/2=3/2 (i.e. [75]). As expected, this is the same value as determined above.

d)   To find E[Y|X=1] the same approach is used as was for the two cases above. However, this time the conditioning variable is X instead of Y. Using [148].

$$f_X(x) = \int_0^3 \frac{1}{5}dy = \frac{3}{5}$$

$$f_{Y|X}(Y \le y|X=1) = \frac{f_{XY}(x, y)}{f_X(x)} = \frac{\frac{1}{5}}{\frac{3}{5}} = \frac{1}{3}, \text{ then adapting [187]:}$$

$$E(Y \le y|X=1) = \int_0^3 yf_{Y|X}(Y \le y|X=x)dy = \frac{1}{3}\left(\frac{y^2}{2}\bigg|_0^3\right) = \frac{1}{3}\left(\frac{9}{2}\right) = \frac{3}{2}$$

As a check, looking into the Y axis at the cross section formed by X=1 we would see the following:

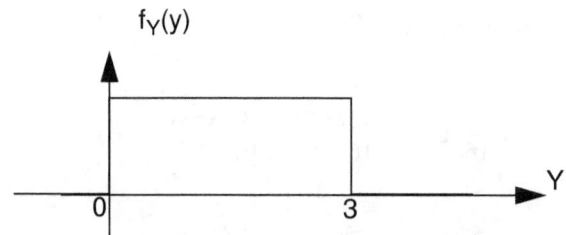

Figure 81 pdf of X when X=1

From Figure 81 we can then calculate the expected value of Y when X=1 by inspection as (3+0)/2=3/2 (i.e. [75]). As expected, this is the same value as determined above.

7   To   determine   if   X   and   Y   are   uncorrelated   we   use   [205], $E[XY] = E[X]E[Y]$. If [205] holds then X and Y are uncorrelated.

$$E[XY] = \int_{-1}^1 \int_{-1}^1 \frac{15}{32}xy(x^2 + y^4)dxdy = \frac{15}{32}\int_{-1}^1 \left(\frac{x^4}{4}y + \frac{x^2}{2}y^5\bigg|_{-1}^1\right)dy$$

$$= \frac{15}{32}\int_{-1}^1 (0)dy = 0$$

Next, to find E[X] and E[Y] we first need to calculate the marginal pdfs using [148] and [149]:

$$f_X(x) = \int_{-\infty}^{\infty} f_{XY}(x, y)dy = \int_{-1}^{1} \frac{15}{32}(x^2 + y^4)dy = \frac{15}{32}\left(x^2 y + \frac{y^5}{5} \bigg|_{-1}^{1} \right)$$

$$= \frac{15}{32}\left(2x^2 + \frac{2}{5}\right) = \frac{15}{16}x^2 + \frac{3}{16}$$

$$f_Y(y) = \int_{-\infty}^{\infty} f_{XY}(x, y)dx = \int_{-1}^{1} \frac{15}{32}(x^2 + y^4)dx = \frac{15}{32}\left(\frac{x^3}{3} + xy^4 \bigg|_{-1}^{1} \right)$$

$$= \frac{15}{32}\left(2y^4 + \frac{2}{3}\right) = \frac{15}{16}y^4 + \frac{5}{16}$$

Now we can calculate E[X] and E[Y]:

$$E[X] = \int_{-1}^{1} x\left(\frac{15}{16}x^2 + \frac{3}{16}\right)dx = \frac{15}{16}\left(\frac{x^4}{4}\right) + \frac{3}{16}\left(\frac{x^2}{2}\right) \bigg|_{-1}^{1} = 0$$

$$E[Y] = \int_{-1}^{1} y\left(\frac{15}{16}y^4 + \frac{5}{16}\right)dy = \frac{15}{16}\left(\frac{y^6}{6}\right) + \frac{3}{16}\left(\frac{y^2}{2}\right) \bigg|_{-1}^{1} = 0$$

Since $E[XY] = E[X]E[Y]$ X and Y are uncorrelated.

To prove (or disprove) independence we apply [157]:

$$\frac{15}{32}(x^2 + y^4) \neq \left(\frac{15}{16}x^2 + \frac{3}{16}\right)\left(\frac{15}{16}y^4 + \frac{5}{16}\right)$$

Since [157] does not hold, X and Y are not statistically independent. ___**This example shows that while independent random variables are always uncorrelated, not all uncorrelated random variables are independent.**___

8 We are given two random variables, X and Y that are uniformly distributed from 2 to 3 and from 1 to 4 respectively, and wish to find the pdf of their sum.

Since we are looking to find the sum of two independent random variables we can apply [216][1]. We 1st form Z=X+Y. Thus the range of Z is from 2+1=3 to 3+4=7.

$f_X(x) = 1$ $2 \leq x \leq 3$ and $f_Y(y) = \frac{1}{3}$ $1 \leq y \leq 4$. Next we graphically determine the limits of integration for [216].

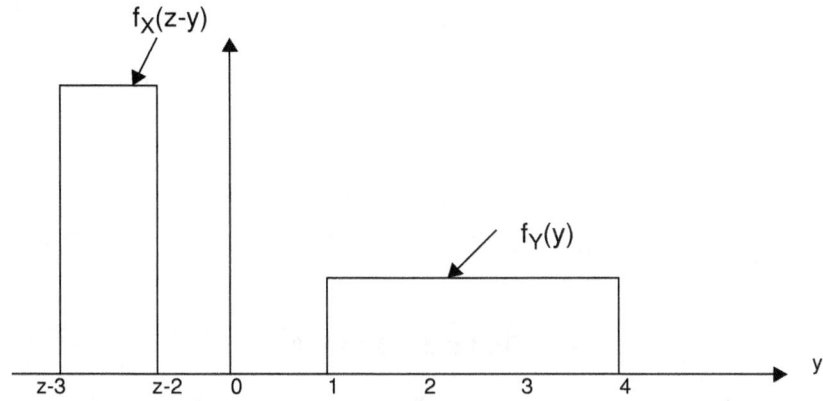

Figure 82 z<3

Figure 82 above shows the condition for z<3. In this case $f_Y(y)f_X(z-y)=0$. Therefore, $f_Z(z)=0$ for z<3.

---

1. Remember [216] can only be used for finding the pdf of the sum of two independent random variables.

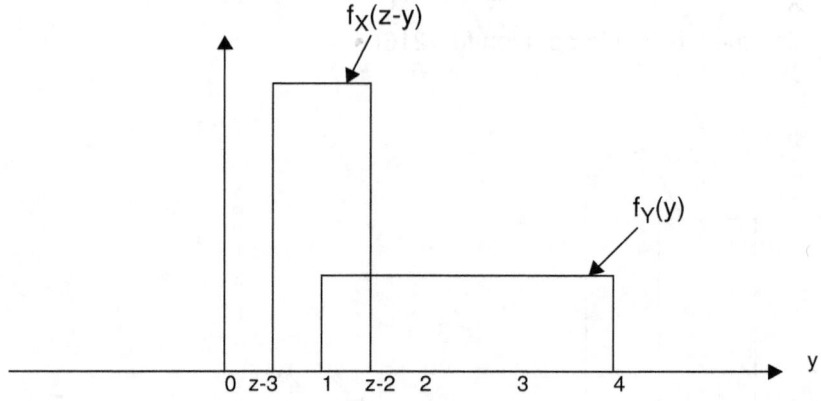

Figure 83 $3 \le z \le 4$

Figure 83 above shows the condition for $3 \le z \le 4$. In this case $f_Y(y)f_X(z-y)=1/3$. Therefore:

$$f_Z(z) = \int_1^{z-2} \frac{1}{3}dy = \frac{z-3}{3} \quad 3 \le z \le 4$$

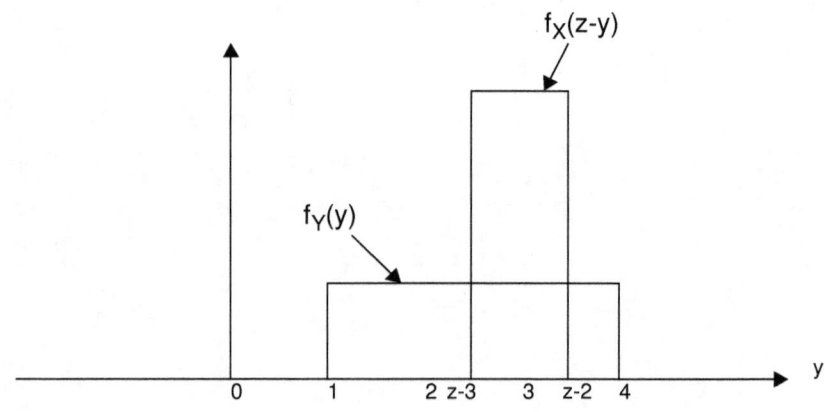

Figure 84 $4 \le z \le 6$

154

Figure 84 above shows the condition for $4 \le z \le 6$. In this case $f_Y(y)f_X(z-y)=1/3$. Therefore:

$$f_Z(z) = \int_{z-3}^{z-2} \frac{1}{3}dy = \frac{1}{3} \quad 4 \le z \le 6$$

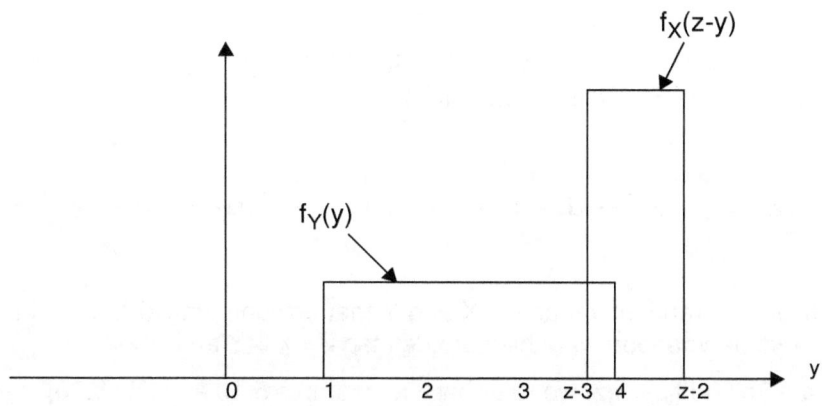

Figure 85 $6 \le z \le 7$

Figure 85 above shows the condition for $6 \le z \le 7$. In this case $f_Y(y)f_X(z-y)=1/3$. Therefore:

$$f_Z(z) = \int_{z-3}^{4} \frac{1}{3}dy = \frac{7-z}{3} \quad 6 \le z \le 7$$

For z>7 $f_Y(y)f_X(z-y)=0$, thus $f_Z(z)=0$.

To summarize:

$$f_Z(z) = \begin{cases} 0 & z < 3 \\ \dfrac{z-3}{3} & 3 \le z < 4 \\ \dfrac{1}{3} & 4 \le z \le 6 \\ \dfrac{7-z}{3} & 6 \le z \le 7 \\ 0 & z > 7 \end{cases}$$

As a check, we know from [169] that E[Z] = E[X]+E[Y] = 2.5 + 2.5 = 5. Now finding E[Z] with the pdf gives:

$$E[Z] = \int_3^4 z\left(\frac{z-3}{3}\right)dz + \int_4^6 z\left(\frac{1}{3}\right)dz + \int_6^7 z\left(\frac{7-z}{3}\right)dz = \frac{11}{18} + \frac{60}{18} + \frac{19}{18} = 5$$

9 Given two random variables, X and Y that are uniformly distributed from 0 to 1. Two new random variables are defined Z= X + Y and W= X:

a) To find $f_{WZ}(w,z)$ we refer back to example 5.19 and use [220].

$$f_{WZ}(w, z) = f_{XY}(k(w, z), m(w, z))\left|\frac{\partial(x, y)}{\partial(w, z)}\right| = f_{XY}(w, z-w) = 1.$$ However, this is only the magnitude of the pdf. What is its shape? Note that:

$$f_{XY}(x, y) = 1[U(x) - U(x-1)][U(y) - U(y-1)]$$

Therefore, the answer is:

$$f_{WZ}(w, z) = f_{XY}(w, z-w) = 1[U(w) - U(w-1)][U(z-w) - U(z-w-1)]$$

b) To find the marginal pdf for z (i.e $f_Z(z)$), [148] can be used as shown below:

$$f_Z(z) = \int_{-\infty}^{\infty} f_{WZ}(w, z)dw$$

However, we need 1st to determine the limits of integration. $[U(w) - U(w-1)]$ and $[U(z-w) - U(z-w-1)]$ can be plotted as:

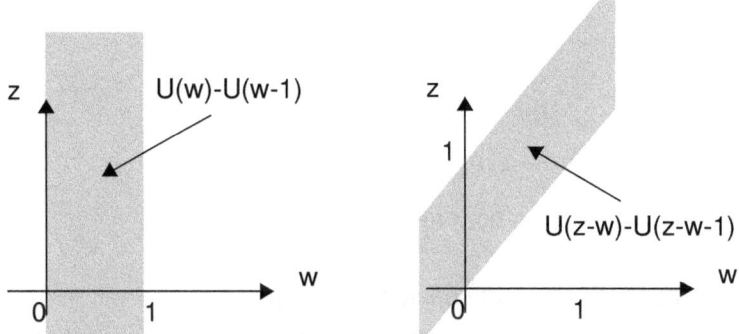

Figure 86 Separate Components of $f_{WZ}(w,z)$

Now multiplying the two components in Figure 86 gives:

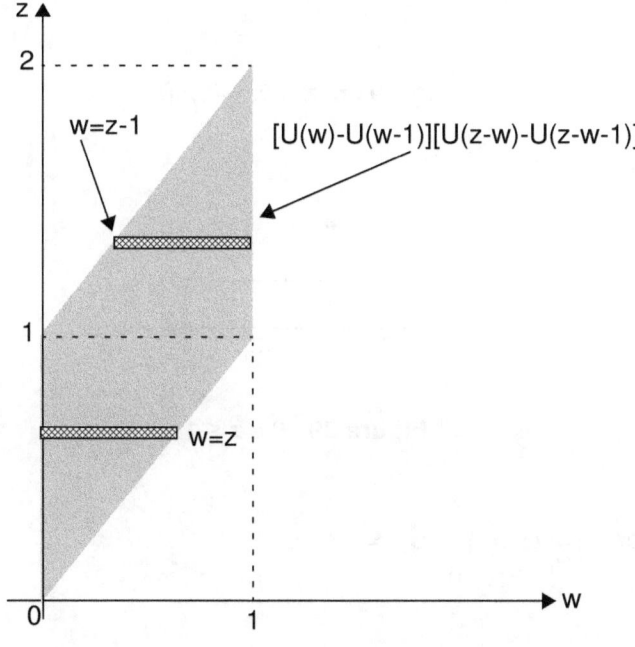

Figure 87 $[U(w)-U(w-1)][U(z-w)-U(z-w-1)]$

Now we can see how to make our limits:

$$f_Z(z) = \int_0^z 1dw = z \quad 0 \le z \le 1$$

$$f_Z(z) = \int_{(z-1)}^1 1dw = 2 + -z \quad 1 \le z \le 2$$

$f_Z(z)$ of course equals 0 for all other values of z.

Now, as an independent check we apply [216][1]:

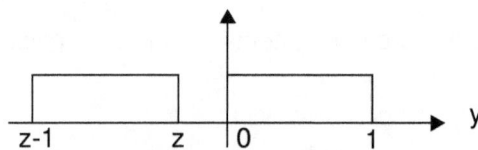

Figure 88 z < 0, $f_Z(z)$=0

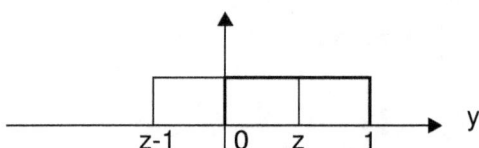

Figure 89 $0 \le z \le 1$

Therefore, $f_Z(z) = \int_0^z 1dy = z \quad 0 \le z \le 1.$

---

1. Even though not explicitly stated, X and Y are actually independent since $f_{XY}(x,y)=f_X(x)f_Y(y)$. Otherwise, we would not be able to apply [216] in this case.

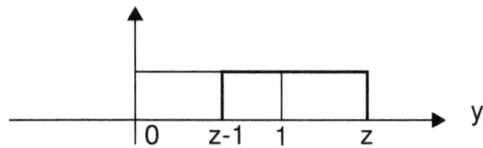

Figure 90 $1 \le z \le 2$

Therefore, $f_Z(z) = \displaystyle\int_{z-1}^{1} 1 \, dy = 2 - z \quad 1 \le z \le 2$.

For z>2 $f_Z(z)$=0. Thus, the results above using [216] match those obtained by [220].

10 To find the joint characteristic function for example 5.1, [233] is used:

$$\Phi_{XY}(\omega_1, \omega_2) = e^{j\omega_1 0 + j\omega_2 0} P(X = 0 \cap Y = 0)$$

$$+ e^{j\omega_1 0 + j\omega_2 1} P(X = 0 \cap Y = 1) + e^{j\omega_1 0 + j\omega_2 2} P(X = 0 \cap Y = 2)$$

$$+ e^{j\omega_1 1 + j\omega_2 0} P(X = 1 \cap Y = 0) + e^{j\omega_1 1 + j\omega_2 1} P(X = 1 \cap Y = 1)$$

$$+ e^{j\omega_1 1 + j\omega_2 2} P(X = 1 \cap Y = 2) = \frac{1}{8} e^{j\omega_1 0 + j\omega_2 0} + \frac{1}{4} e^{j\omega_1 0 + j\omega_2 1}$$

$$+ \frac{1}{8} e^{j\omega_1 0 + j\omega_2 2} + \frac{1}{8} e^{j\omega_1 1 + j\omega_2 0} + \frac{1}{4} e^{j\omega_1 1 + j\omega_2 1} + \frac{1}{8} e^{j\omega_1 1 + j\omega_2 2}$$

As a check, we apply [230] to the characteristic function to find $E[XY^2]$ which was calculated in example 5.9:

$$\frac{1}{j^3}\frac{\partial^3}{\partial\omega_1^1\partial\omega_2^2}\Phi_{XY}(\omega_1,\omega_2)\Big|_{\omega_1=\omega_2=0}$$

$$= \frac{1}{j^3}\left(\frac{j^3}{4}e^{j0(1)}+j0(1)+\frac{4j^3}{8}e^{j0(1)}+j0(2)\right) = \frac{3}{4}$$

This value matches the value calculated in example 5.9 for $E[XY^2]$.

Chapter 6   Additional Topics

In this chapter we introduce some additional topics to round out our study of multiple random variables.

6.1   Correlation Coefficients

The definition of the correlation coefficient between two random variables is the ratio of the covariance to the product of the standard deviations of the two random variables as follows:

[234] $\rho_{XY} = \dfrac{\sigma_{XY}}{\sigma_X \sigma_Y} = \dfrac{E[XY] - E[X]E[Y]}{\sigma_X \sigma_Y}$

The correlation coefficient gives a measure of how much two random variables are linearly dependent upon each other. Furthermore:

[235] $-1 \le \rho_{XY} \le 1$

When the correlation coefficient of two random variables is equal to 0, the random variables are said to be uncorrelated.

The following example will try to illustrate the correlation coefficient.

Example 6.1:

Calculate ρ_{XY} for the following cases:

a)   Y = X
b)   Y = -X
c)   Y = X^2
d)   Y and X are independent.
a)   Setting Y = X in [234] gives:

$$\rho_{XY} = \frac{E[XX] - E[X]E[X]}{\sigma_X \sigma_X} = \frac{E[X^2] - (E[X])^2}{\sigma_X^2} = \frac{\sigma_X^2}{\sigma_X^2} = 1$$

Thus, it is seen that if X and Y are linearly related to each other, then $\rho_{XY} = 1$.

b)   Setting Y = -X in [234] gives:

$$\rho_{XY} = \frac{E[X(-X)] - E[X]E[-X]}{\sigma_X \sigma_X} = \frac{E[-X^2] + (E[X])^2}{\sigma_X^2} = \frac{-\sigma_X^2}{\sigma_X^2} = -1$$

Thus, it is seen that if X and Y are linearly but inversely related to each other, then $\rho_{XY} = -1$.

c) Setting $Y = X^2$ in [234] gives:

$$\rho_{XY} = \frac{E[X^3] - E[X]E[X^2]}{\sigma_X \sigma_{X^2}} \rightarrow -1 \leq \rho_{XY} \leq 1$$

The actual value of ρ_{XY} depends upon the distributions and standard deviations of X and Y. Thus, if X and Y are related, but not linearly related, the correlation coefficient will be between -1 and 1. The more linearly related X and Y are, then the closer to 1 (or -1) the correlation coefficient will be.

d) If X and Y are independent, then [234] gives:

$$\rho_{XY} = \frac{E[X]E[Y] - E[X]E[Y]}{\sigma_X \sigma_Y} = 0$$

Thus, a value of 0 for the correlation coefficient shows that X and Y are not linearly related. However, note that while independent random variables have a correlation coefficient of 0, a correlation coefficient of 0 does not necessarily imply independence. ∎

Consider next the sum of two random variables, Z=X+Y. Calculating the variance:

$$\sigma^2_Z = E[(Z - \bar{Z})^2] = E[(X - \bar{X} + Y - \bar{Y})^2]$$

$$= E[X^2 - X\bar{X} + XY - X\bar{Y} - \bar{X}X + \bar{X}^2 - \bar{X}Y + \bar{X}\bar{Y} + YX - Y\bar{X} + Y^2 - Y\bar{Y}$$

$$-\bar{Y}X + \bar{Y}\bar{X} - \bar{Y}Y + \bar{Y}^2]$$

Taking the expected values of each term gives:

$$\sigma^2_Z = \overline{X^2} - \overline{X}^2 + \overline{XY} - \overline{X}\overline{Y} - \overline{X}^2 + \overline{X}^2 - \overline{X}\overline{Y} + \overline{X}\overline{Y} + \overline{XY} - \overline{X}\overline{Y} + \overline{Y^2} - \overline{Y}^2$$

$$-\overline{Y}\overline{X} + \overline{Y}\overline{X} - \overline{Y}^2 + \overline{Y}^2]$$

Collecting terms gives:

[236] $\sigma^2_Z = \overline{X^2} - \overline{X}^2 + 2\overline{XY} - 2\overline{X}\overline{Y} + \overline{Y^2} - \overline{Y}^2 = \sigma^2_X + 2\sigma_X\sigma_Y\rho_{XY} + \sigma^2_Y$

If X and Y are independent or uncorrelated, then the correlation coefficient reduces to 0 and we have:

[237] $\sigma^2_Z = \sigma^2_X + \sigma^2_Y$

Thus, the variance of the sum of two independent or uncorrelated random variables is the sum of their individual variances. Remember that this result only applies to random variables which are independent or uncorrelated. If the random variables are neither independent or uncorrelated, then [236] must be used.

6.2   Jointly Gaussian Random Variables

The representation of an N-variate Gaussian random variable can be very unwieldy to represent, even for N=2. To help ease this problem, matrix notation is used. The representation of an N-variate Gaussian random variable using this matrix notation is given as[1]:

[238] $f_{X_1X_2...X_N}(x_1, x_2, ..., x_N) = \dfrac{\sqrt{|C^{-1}|}}{(2\pi)^{N/2}} e^{-\left(\frac{X^TC^{-1}X}{2}\right)}$,

where:

[239] $X = \begin{bmatrix} x_1 - \overline{X_1} \\ x_2 - \overline{X_2} \\ ... \\ ... \\ x_N - \overline{X_N} \end{bmatrix}$

---

1. Note that bold capital letters are being used to represent matrices.

$$
[240]\ \ \mathbf{C} = \begin{bmatrix} \sigma^2_{X_1} & \rho_{X_1 X_2}\sigma_{X_1}\sigma_{X_2} & \cdots\ \cdots & \rho_{X_1 X_N}\sigma_{X_1}\sigma_{X_N} \\ \rho_{X_2 X_1}\sigma_{X_2}\sigma_{X_1} & \sigma^2_{X_2} & \cdots\ \cdots & \rho_{X_2 X_N}\sigma_{X_2}\sigma_{X_N} \\ \cdots & \cdots & \cdots\ \cdots & \cdots \\ \cdots & \cdots & \cdots\ \cdots & \cdots \\ \rho_{X_N X_1}\sigma_{X_N}\sigma_{X_1} & \cdots & \cdots\ \cdots & \sigma^2_{X_N} \end{bmatrix}
$$

Example 6.2:

Use [238] to determine the joint pdf of two Gaussian random variables, X_1 and X_2.

For N=2, [238], [239] and [240] become:

$$
[241]\ \ f_{X_1 X_2}(x_1, x_2) = \frac{\sqrt{|\mathbf{C}^{-1}|}}{2\pi}e^{-\left(\frac{\mathbf{X}^T \mathbf{C}^{-1}\mathbf{X}}{2}\right)}
$$

$$
\mathbf{X} = \begin{bmatrix} x_1 - \overline{X_1} \\ x_2 - \overline{X_2} \end{bmatrix} \qquad \mathbf{C} = \begin{bmatrix} \sigma^2_{X_1} & \rho_{X_1 X_2}\sigma_{X_1}\sigma_{X_2} \\ \rho_{X_2 X_1}\sigma_{X_2}\sigma_{X_1} & \sigma^2_{X_2} \end{bmatrix}
$$

$\mathbf{C}$ is called the covariance matrix. Then, using the matrix inversion formula[1]:

$$
\begin{bmatrix} a & b \\ c & d \end{bmatrix}^{-1} = \begin{bmatrix} \dfrac{d}{ad-bc} & \dfrac{-b}{ad-bc} \\ \dfrac{-c}{ad-bc} & \dfrac{a}{ad-bc} \end{bmatrix},
$$

---

1. Chapters 1 and 2 of reference [12] give instruction on basic matrix operations.

$$C^{-1} = \begin{bmatrix} \dfrac{\sigma^2 x_2}{\sigma^2 x_2 \sigma^2 x_1 \left(1 - \left(\rho_{x_1 x_2}\right)^2\right)} & -\dfrac{\rho_{x_1 x_2} \sigma_{x_1} \sigma_{x_2}}{\sigma^2 x_2 \sigma^2 x_1 \left(1 - \left(\rho_{x_1 x_2}\right)^2\right)} \\[3ex] -\dfrac{\rho_{x_1 x_2} \sigma_{x_1} \sigma_{x_2}}{\sigma^2 x_2 \sigma^2 x_1 \left(1 - \left(\rho_{x_1 x_2}\right)^2\right)} & \dfrac{\sigma^2 x_1}{\sigma^2 x_2 \sigma^2 x_1 \left(1 - \left(\rho_{x_1 x_2}\right)^2\right)} \end{bmatrix}$$

and then taking the determinant[1]:

$$|C^{-1}| = \dfrac{\sigma^2 x_2 \sigma^2 x_1 \left(1 - \left(\rho_{x_1 x_2}\right)^2\right)}{\left[\sigma^2 x_2 \sigma^2 x_1 \left(1 - \left(\rho_{x_1 x_2}\right)^2\right)\right]^2} = \dfrac{1}{\sigma^2 x_2 \sigma^2 x_1 \left(1 - \left(\rho_{x_1 x_2}\right)^2\right)}$$

and finally, taking the square root:

[242] $\sqrt{|C^{-1}|} = \dfrac{1}{\sqrt{\sigma^2 x_2 \sigma^2 x_1 \left(1 - \left(\rho_{x_1 x_2}\right)^2\right)}} = \dfrac{1}{\sigma_{x_1} \sigma_{x_2} \sqrt{1 - \left(\rho_{x_1 x_2}\right)^2}}$

Next we calculate $X^T C^{-1} X$ using the formula[2]:

[243] $X^T C^{-1} X = \begin{bmatrix} a \\ b \end{bmatrix}^T \begin{bmatrix} \dfrac{c}{\Delta} & -\dfrac{d}{\Delta} \\[2ex] -\dfrac{e}{\Delta} & \dfrac{f}{\Delta} \end{bmatrix} \begin{bmatrix} a \\ b \end{bmatrix} = \dfrac{ca^2 + fb^2 - ab(d + e)}{\Delta}$, where

$\Delta = \sigma^2 x_2 \sigma^2 x_1 \left(1 - \left(\rho_{x_1 x_2}\right)^2\right)$, $\quad a = x_1 - \overline{X_1}$, $\quad b = x_2 - \overline{X_2}$,

$c = \sigma^2 x_2$, $d = e = \rho_{x_1 x_2} \sigma_{x_1} \sigma_{x_2}$, and $f = \sigma^2 x_1$.

Inserting these values into [243] gives:

---

1. Note that $\rho_{x_1 x_2} = \rho_{x_2 x_1}$.

2. This formula was derived using a math program.

$$[244] \quad \frac{\sigma^2_{X_2}(x_1 - \overline{X_1})^2 + \sigma^2_{X_1}(x_2 - \overline{X_2})^2 - (x_1 - \overline{X_1})(x_2 - \overline{X_2})\left(2\rho_{X_1 X_2}\sigma_{X_1}\sigma_{X_2}\right)}{\sigma^2_{X_2}\sigma^2_{X_1}\left(1 - \left(\rho_{X_1 X_2}\right)^2\right)}$$

$$= \frac{\left[\frac{(x_1 - \overline{X_1})}{\sigma_{X_1}}\right]^2 + \left[\frac{(x_2 - \overline{X_2})}{\sigma_{X_2}}\right]^2 - \left(\frac{x_1 - \overline{X_1}}{\sigma_{X_1}}\right)\left(\frac{x_2 - \overline{X_2}}{\sigma_{X_2}}\right)\left(2\rho_{X_1 X_2}\right)}{\left(1 - \left(\rho_{X_1 X_2}\right)^2\right)}$$

Thus, inserting [243] and [244] into [241] results in the bivariate Gaussian distribution:

$$[245] \quad f_{X_1 X_2}(x_1, x_2) = \frac{e^{-\frac{\left[\frac{(x_1 - \overline{X_1})}{\sigma_{X_1}}\right]^2 + \left[\frac{(x_2 - \overline{X_2})}{\sigma_{X_2}}\right]^2 - \left(\frac{x_1 - \overline{X_1}}{\sigma_{X_1}}\right)\left(\frac{x_2 - \overline{X_2}}{\sigma_{X_2}}\right)\left(2\rho_{X_1 X_2}\right)}{2\left(1 - \left(\rho_{X_1 X_2}\right)^2\right)}}}{2\pi\sigma_{X_1}\sigma_{X_2}\sqrt{1 - \left(\rho_{X_1 X_2}\right)^2}}$$

Notice how complex this pdf is for just two Gaussian random variables. Thus, the necessity of being able to represent the pdf of N-variate Gaussian random variables by [238] can be better appreciated. ∎

6.3 Mean Square Estimation

In this section we present two methods for estimating the value of an unobservable random variable, Y, given that we can observe another random variable X. Let's say that Y can be estimated as a function of X called g(X) as follows:

$$\hat{Y} = g(X)$$

Therefore, the mean square error in the estimation is:

$$[246] \quad E[(Y - \hat{Y})^2] = E[(Y - g(X))^2]$$

How to select g(X) so that the mean square error is the lowest that it can be?

Inserting [246] into [162] results in:

[247] $E[(Y - \hat{Y})^2] = \displaystyle\int_{-\infty}^{\infty} \int_{-\infty}^{\infty} (y - g(x))^2 f_{XY}(x, y) dx dy$

Furthermore, using the result of [181] in [247] gives:

[248] $E[(Y - \hat{Y})^2] = \displaystyle\int_{-\infty}^{\infty} \int_{-\infty}^{\infty} (y - g(x))^2 f_{Y|X}(Y \leq y | X = x) f_X(x) dx dy$

Rearranging the terms:

[249] $E[(Y - \hat{Y})^2] = \displaystyle\int_{-\infty}^{\infty} f_X(x) \left[\int_{-\infty}^{\infty} (y - g(x))^2 f_{Y|X}(Y \leq y | X = x) dy \right] dx$

We can see that [249] is a minimum for all x if:

$\displaystyle\int_{-\infty}^{\infty} (y - g(x))^2 f_{Y|X}(Y \leq y | X = x) dy$ is a minimum.

To find the minimum we use basic calculus and take the derivative using Leibnitz's Rule[1] and set it to 0:

$\dfrac{\partial}{\partial g(x)} \left[\displaystyle\int_{-\infty}^{\infty} (y - g(x))^2 f_{Y|X}(Y \leq y | X = x) dy \right] = 0$

$-2 \displaystyle\int_{-\infty}^{\infty} (y - g(x)) f_{Y|X}(Y \leq y | X = x) dy = 0$ Thus,

---

1. $\dfrac{\partial}{\partial \lambda} \displaystyle\int_a^b f(x, \lambda) dx = \int_a^b \dfrac{\partial}{\partial \lambda} f(x, \lambda) dx$.

$$\int_{-\infty}^{\infty} y f_{Y|X}(Y \le y|X=x)dy = \int_{-\infty}^{\infty} g(x) f_{Y|X}(Y \le y|X=x)dy$$

and simplifying further:

$$E[Y \le y|X=x] = g(x)\int_{-\infty}^{\infty} f_{Y|X}(Y \le y|X=x)dy = g(x)$$

Thus, we finally arrive at the value of g(X) which minimizes the mean square error of [246]:

[250] $\hat{Y} = g(X) = E[Y \le y|X=x]$

Example 6.3:

Given Y=3X+4. $f_{XY}(x,y)=(x+y)/8$ for x,y ranging from 0 to 2, and the marginal pdf of x is $f_X(x)=(x+1)/4$. What is the mean square estimator of Y and what is the resulting mean square error?

$$\hat{Y} = g(X) = E[3X+4 \le y|X=x] = \int_0^2 y\frac{f_{XY}(y,x)}{f_X(x)}dy$$

$$= \int_0^2 (3x+4)\frac{(x+y)}{2(x+1)}dy = \frac{(3x+4)}{2(x+1)}\int_0^2 x+y dx = \frac{(3x+4)}{2(x+1)}\left[xy+\frac{y^2}{2}\right]\Big|_0^2$$

$$= \frac{(3x+4)}{2(x+1)}[2x+2] = 3x+4$$

Now the minimum mean square error is found to be:

$$E[(Y-\hat{Y})^2] = E[(3X+4-3X-4)^2] = 0 \blacksquare$$

Before we look at using a linear function as an estimator, let us examine estimating Y by a constant: $\hat{Y} = c$.

Then the mean square error is:

168

[251] $E[(Y - \hat{Y})^2] = E[(Y - c)^2] = \int_{-\infty}^{\infty} (y - c)^2 f_Y(y) dy$

To find the value of c that minimizes the mean square error, we take the derivative of the integral using Leibnitz's Rule[1], set it to 0 and solve for c:

$$\frac{\partial}{\partial c} \int_{-\infty}^{\infty} (y - c)^2 f_Y(y) dy = \int_{-\infty}^{\infty} \frac{\partial}{\partial c} (y - c)^2 f_Y(y) dy$$

$$= -2 \int_{-\infty}^{\infty} (y - c) f_Y(y) dy = 0 \rightarrow c \int_{-\infty}^{\infty} f_Y(y) dy = \int_{-\infty}^{\infty} y f_Y(y) dy$$

This finally reduces to:

[252] $c = E[Y]$

Thus, if estimating Y by a constant, using $\hat{Y} = E[Y]$ will minimize the mean square error. Keep [252] in the back of your mind as we will utilize it shortly.

Next, we present an estimation method which uses a linear function estimator as shown below:

[253] $\hat{Y} = aX + b$

This method does not generally provide as good an estimation as [250], but may be simpler in some cases. The mean square error using [253] is:

[254] $E[(Y - \hat{Y})^2] = E[(Y - aX - b)^2]$

We can write [254] as:

[255] $E[((Y - aX) - b)^2]$

Comparing to [251], we can say that the value of b which minimizes the mean square error is $b = E[Y - aX]$. Inserting this value into [255] results in:

---

1. $\dfrac{\partial}{\partial \lambda} \displaystyle\int_{a}^{b} f(x, \lambda) dx = \int_{a}^{b} \dfrac{\partial}{\partial \lambda} f(x, \lambda) dx$.

$$E[((Y-aX)-b)^2] = E[((Y-aX)-E[Y-aX])^2] = E[((Y-\overline{Y})-a(X-\overline{X}))^2]$$

$$= E[(Y-\overline{Y})^2 - 2a(X-\overline{X})(Y-\overline{Y}) + a^2(X-\overline{X})^2]$$

$$= \sigma_Y^2 - 2a\rho_{XY}\sigma_X\sigma_Y + a^2\sigma_X$$

Taking the derivative of the above result, setting it to 0, and solving for a, results in the value of a, which will provide the minimum mean square error:

$$\frac{\partial}{\partial a}(\sigma_Y^2 - 2a\rho_{XY}\sigma_X\sigma_Y + a^2\sigma_X^2) = -2\rho_{XY}\sigma_X\sigma_Y + 2a\sigma_X^2 = 0 \rightarrow a = \frac{\rho_{XY}\sigma_Y}{\sigma_X}$$

So to summarize, if $\hat{Y} = aX + b$ is used as an estimator for Y, then the values of a and b that minimize the mean square error are $a = \dfrac{\rho_{XY}\sigma_Y}{\sigma_X}$ and

$b = E[Y - aX] = \overline{Y} - a\overline{X}$.

6.4   The Central Limit Theorem

The central limit theorem (clt), basically states that the cdf of a sum of n independent random variables that are identically distributed[1], and appropriately normalized, will become approximately Gaussian as n increases. This can be very useful for calculations where there are a large sum of random variables. The normalized sum is given by:

$$[256] \quad Z_n = \frac{X_1 + X_2 + \ldots X_n - n\overline{X}}{\sigma_X\sqrt{n}}$$

Now as n goes to infinity, the cdf of Z_n will approach to being Gaussian as follows:

$$[257] \quad \lim_{n \to \infty} F_{Z_n}(z) = \lim_{n \to \infty} P(Z_n \le z) = G(z)$$

Then G(z) can be looked up in Table 2 as shown in the following example.

Example 6.4:

---

1. This implies that the random variables have identical means and variances.

The total random noise current in a circuit consists of the sum of ten independent and identically distributed random noise currents that have a mean of 0.04amps and a variance of 0.007amps^2. What is the probability that the total random noise current will be ≤ 0.350amps?

Applying [256]:

$$Z_n = \frac{X_1 + X_2 + ...X_n - n\overline{X}}{\sigma_X \sqrt{n}} = \frac{0.350 - 10(0.04)}{\sqrt{(0.007)(10)}} = -0.188982$$

Next using [80] and Table 2:

$$G(-z) = 1 - G(z) = 1 - G(0.188982) \cong 0.5753$$

One could only imagine the effort in developing the cdf or pdf[1]for the total random current and then having to do another calculation to find the desired probability. This should be evidence enough of the clt's value. ∎

It is worth noting that the value of n to get a good approximation is difficult to predict.

6.5 Exercises

1 Find the correlation coefficient for the two cases below. Comment on the results. X is uniformly distributed from 0 to 1.

a) $Y = X^{1.1}$

b) $Y = X^{1.5}$

2 The random variables X and Y have the following distribution: $f_{XY}(x,y) = 6x^2y$ where $0<x<1$ and $0<y<1$. Find the variance of their sum.

3 What is the variance of aX +bY? X and Y are independent and a and b are constants.

4 Consider two uncorrelated Gaussian random variables. Determine their joint pdf and compare with the corresponding pdf of two independent Gaussian random variables. Comment on the results.

---

1. 10 convolutions would be needed in this case. You would also need to know the pdfs as well, but with the clt you just need the mean and variance (assuming that the random variables are independent and identically distributed).

5 Rework example 6.4 but this time find the probability that the total random current is > 0.420 amps.

6 If Y is estimated using $\hat{Y} = aX + b$, what is the minimum mean square error?

6.6   Answers to Exercises

1

a) $Y = X^{1.1}$

Using [234]:

$$\rho_{XY} = \frac{\sigma_{XY}}{\sigma_X \sigma_Y} = \frac{E[X^{2.1}] - E[X]E[X^{1.1}]}{\sigma_X \sigma_{X^{1.1}}}$$

$$E[X^{2.1}] = \int_0^1 x^{2.1} dx = \left.\frac{x^{3.1}}{3.1}\right|_0^1 = 0.32258$$

$$E[X^{1.1}] = \int_0^1 x^{1.1} dx = \left.\frac{x^{2.1}}{2.1}\right|_0^1 = 0.47619$$

$$E[X] = \int_0^1 x dx = \left.\frac{x^2}{2}\right|_0^1 = \frac{1}{2}$$

$$\sigma_X^2 = \int_0^1 \left(x - \frac{1}{2}\right)^2 dx = \frac{1}{12} \to \sigma_X = \frac{1}{\sqrt{12}}$$

$$\sigma_{X^{1.1}}^2 = \int_0^1 (x^{1.1} - 0.47619)^2 dx = 0.08574 \to \sigma_{X^{1.1}} = 0.29282$$

$$\rho_{XY} = \frac{E[X^{2.1}] - E[X]E[X^{1.1}]}{\sigma_X \sigma_{X^{1.1}}} = \frac{(0.32258) - (0.5)(0.47619)}{\left(\frac{1}{\sqrt{12}}\right)(0.29282)} = 0.99944$$

b) $Y = X^{1.5}$

$$E[X^{2.5}] = \int_0^1 X^{2.5}dx = \frac{X^{3.5}}{3.5}\Big|_0^1 = 0.28571$$

$$E[X^{1.5}] = \int_0^1 X^{1.5}dx = \frac{X^{2.5}}{2.5}\Big|_0^1 = 0.4$$

$$\sigma^2_{X^{1.5}} = \int_0^1 (x^{1.5} - 0.4)^2 dx = 0.09 \rightarrow \sigma_{X^{1.5}} = 0.3$$

$$\rho_{XY} = \frac{E[X^{2.5}] - E[X]E[X^{1.5}]}{\sigma_X \sigma_{X^{1.5}}} = \frac{0.28571 - (0.5)(0.4)}{\left(\frac{1}{\sqrt{12}}\right)(0.3)} = 0.98966$$

ρ_{XY} for $X^{1.1}$ = 0.99944 while ρ_{XY} for $X^{1.5}$ = 0.98966[1]. Since $X^{1.1}$ is very close to X it makes sense that the corresponding ρ_{XY} is close to 1. Since $X^{1.5}$ is a little further away from X than is $X^{1.1}$ we expect that its ρ_{XY} would be a little further away from 1 than the ρ_{XY} for $X^{1.1}$ is.

2 X and Y were determined in section 5.13 exercise 4 to be independent so we can use [237] and just sum their variances. Their marginal pdfs were calculated in section 5.13 exercise 4 as:

$$f_X(x) = 3x^2 \text{ and } f_Y(y) = 2y.$$

The means are:

$$E[X] = \int_0^1 3x^3 dx = \frac{3x^4}{4}\Big|_0^1 = \frac{3}{4}$$

$$E[X^2] = \int_0^1 3x^4 dx = \frac{3x^5}{5}\Big|_0^1 = \frac{3}{5}$$

---

1. If rounded off values were not used then ρ_{XY} for $X^{1.1}$ = 0.99948 while ρ_{XY} for $X^{1.5}$ = 0.98974.

$$\sigma_X^2 = \overline{X^2} - \overline{X}^2 = \frac{3}{5} - \left(\frac{3}{4}\right)^2 = \frac{3}{80}$$

$$E[Y] = \int_0^1 2y^2 dy = \left. \frac{2y^3}{3} \right|_0^1 = \frac{2}{3}$$

$$E[Y^2] = \int_0^1 2y^3 dy = \left. \frac{2y^4}{4} \right|_0^1 = \frac{1}{2}$$

$$\sigma_Y^2 = \overline{Y^2} - \overline{Y}^2 = \frac{1}{2} - \left(\frac{2}{3}\right)^2 = \frac{1}{18}$$

Thus:

$$\sigma_{X+Y}^2 = \sigma_X^2 + \sigma_Y^2 = \frac{3}{80} + \frac{1}{18} = \frac{67}{720}$$

3 $Z = aX + bY$ and $\overline{Z} = a\overline{X} + b\overline{Y}$.

$$\sigma_Z^2 = E[(Z - \overline{Z})^2] = E[((aX + bY) - (a\overline{X} + b\overline{Y}))^2]$$

$$\sigma_Z^2 = E[(a(X - \overline{X}) + b(Y - \overline{Y}))^2]$$

$$= E[a^2(X - \overline{X})^2 + 2ab(X - \overline{X})(Y - \overline{Y}) + b^2(Y - \overline{Y})^2]$$

$$= E[a^2(X - \overline{X})^2 + 2ab(XY - \overline{X}Y - X\overline{Y} + \overline{X}\,\overline{Y}) + b^2(Y - \overline{Y})^2]$$

Because of independence we can write the expectation as:

$$\sigma_Z^2 = a^2\sigma_X^2 + 2ab(\overline{X}\,\overline{Y} - \overline{X}\,\overline{Y} - \overline{X}\,\overline{Y} + \overline{X}\,\overline{Y}) + b^2\sigma_Y^2 = a^2\sigma_X^2 + b^2\sigma_Y^2$$

4 The bi-variate Gaussian pdf is given by [245]. If the two random variables are uncorrelated then their correlation coefficient will be equal to 0. So setting the correlation coefficient to 0 in [245] yields:

$$[258] \quad f_{X_1 X_2}(x_1, x_2) = \frac{e^{-\frac{\left(\left[\frac{(x_1 - \overline{X_1})}{\sigma_{X_1}}\right]^2 + \left[\frac{(x_2 - \overline{X_2})}{\sigma_{X_2}}\right]^2 \right)}{2}}}{2\pi\sigma_{X_1}\sigma_{X_2}}$$

For independent Gaussian random variables the bi-variate Gaussian pdf is given by the product of the marginal pdfs.

$$f_{X_1}(x_1) = \frac{e^{-\frac{\left[\frac{(x_1 - \overline{X_1})}{\sigma_{X_1}}\right]^2}{2}}}{\sqrt{2\pi\sigma_{X_1}^2}} \quad \text{and} \quad f_{X_2}(x_2) = \frac{e^{-\frac{\left[\frac{(x_2 - \overline{X_2})}{\sigma_{X_2}}\right]^2}{2}}}{\sqrt{2\pi\sigma_{X_2}^2}}$$

Multiplying the two marginal pdfs together, yields the same result as in [258] which implies that the two uncorrelated Gaussian random variables are indeed independent. This is an exception to rule that uncorrelated random variables are not necessarily independent.

5

Applying [256]:

$$Z_n = \frac{X_1 + X_2 + \dots X_n - n\overline{X}}{\sigma_X \sqrt{n}} = \frac{0.420 - 10(0.04)}{\sqrt{(0.007)(10)}} = 0.07559$$

Next using Table 2:

$$P(0.420 > z) = 1 - G(0.07559) \cong 1 - 0.5299 = 0.4701 \text{ [1]}$$

6 The mean square error is:

---

1. 0.5299 is the average of G(0.07) and G(0.08) from Table 2, i.e. ~G(0.075). Using a math program, the exact value is 0.53013, which is pretty close.

$$E[(Y - aX - b)^2] = E[(Y - aX - \bar{Y} + a\bar{X})^2] = E[((Y - \bar{Y}) - a(X - \bar{X}))^2]$$
$$= E[(Y - \bar{Y})^2 - 2a(X - \bar{X})(Y - \bar{Y}) + a^2(X - \bar{X})^2]$$
$$= \sigma_Y^2 - 2a\overline{XY} + 2a\bar{X}\bar{Y} + 2a\bar{X}\bar{Y} - 2a\bar{X}\bar{Y} + a^2\sigma_X^2$$
$$= \sigma_Y^2 - 2a\rho_{XY}\sigma_X\sigma_Y + a^2\sigma_X^2$$

Inserting a:

$$\sigma_Y^2 - 2\left(\frac{\rho_{XY}\sigma_Y}{\sigma_X}\right)\rho_{XY}\sigma_X\sigma_Y + \left(\frac{\rho_{XY}\sigma_Y}{\sigma_X}\right)^2\sigma_X^2$$
$$= \sigma_Y^2 - 2(\rho_{XY}\sigma_Y)^2 + (\rho_{XY}\sigma_Y)^2 = \sigma_Y^2(1 - \rho_{XY}^2)$$

Chapter 7   Final Quiz

The questions below cover the material presented in chapters 4, 5 and 6. To get the most benefit from these questions, try working them out first before looking at the solutions or the previous chapters. If you absolutely cannot solve a problem, then try looking back over the material in the previous chapters before looking at the solutions (which should be the last resort). These questions are pretty straight forward and should not take more than 2 hours to complete. Calculators are allowed.

7.1   Final Quiz

Q.1 $f_{XYZ}(x,y,z)=kxy^2e^{-z}$ $0<x,y,z<1$. Are X, Y, and Z statistically independent?

Q.2 Find the pdf of Z, given that Y and X are uniformly distributed from -3 to 4 and -1 to 1 respectively. Then find the expected value of Z.

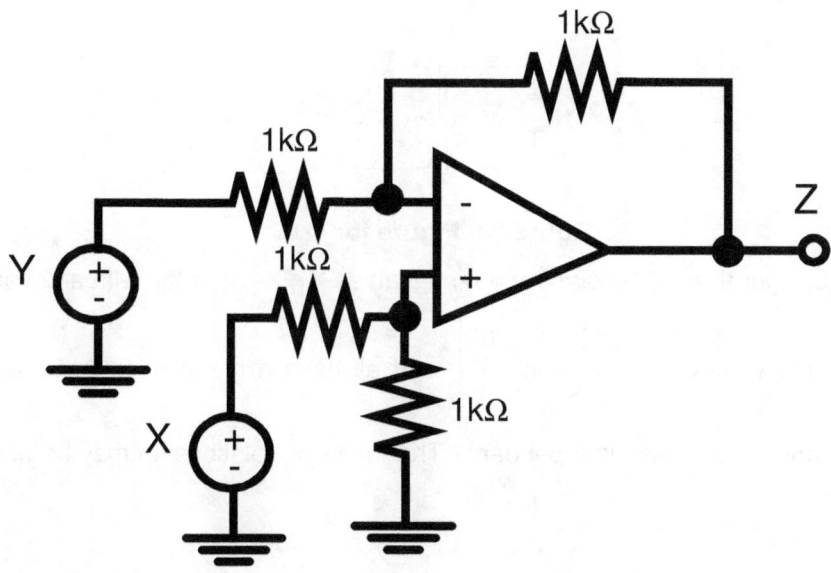

Figure 91 Figure for Q.2

Q.3 Given a Gaussian random variable with a mean of $\overline{X}$ and a standard deviation of σ, find the characteristic function. Then, using the characteristic function, find the mean, mean square value, and the variance.

Q.4 Find the correlation coefficient for Y=3X+2 if X is uniformly distributed from -1 to 0.

Q.5 Given the control system below, find the expected value of Y if X is a zero mean Gaussian random variable with unit variance. A is a constant.

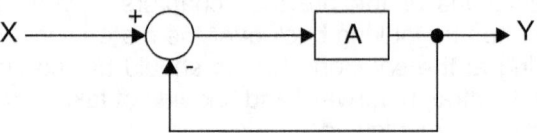

Figure 92 Figure for Q.5

Q.6 X is a zero mean, unit variance, Gaussian random variable. The diode has a forward voltage drop of 0.7volts, but is otherwise ideal. What is the expected value of the voltage Y? If you set up the integral correctly, that is sufficient to give yourself full credit for this problem.

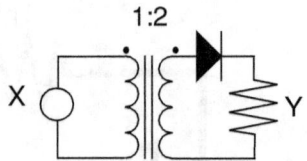

Figure 93 Figure for Q.6

Q.7 Consider three Gaussian random variables X_1, X_2, and X_3, with a covariance matrix of: $\mathbf{C} = \begin{bmatrix} 0.25 & 0 & 0 \\ 0 & 5.00 & 0 \\ 0 & 0 & 0.05 \end{bmatrix}$ and all have a mean of 0. Are these three random variables independent? The following relationship may be useful:

$$\mathbf{C}^{-1} = \begin{bmatrix} a & 0 & 0 \\ 0 & b & 0 \\ 0 & 0 & c \end{bmatrix}^{-1} = \begin{bmatrix} \frac{1}{a} & 0 & 0 \\ 0 & \frac{1}{b} & 0 \\ 0 & 0 & \frac{1}{c} \end{bmatrix}$$

Q.8 Assuming that each current source is independent and identically distributed, find P(I > 1.6A).

178

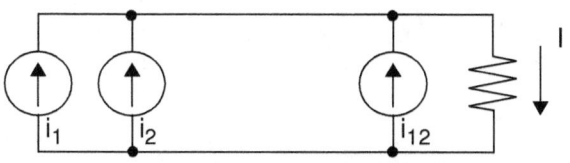

$E[i_n] = 0.12A$, Variance of $i_n = 0.2A$

Figure 94 Figure for Q.8

7.2   Solutions to Final Quiz

Q.1 $f_{XYZ}(x,y,z)=kxy^2e^{-z}$ $0<x,y,z<1$. Are x, y, and z independent?

First let's find k:

$$\int_0^1\int_0^1\int_0^1 kxy^2e^{-z}dxdydz = k\left(\frac{x^2}{2}\Big|_0^1\right)\left(\frac{y^3}{3}\Big|_0^1\right)\left(-e^{-z}\Big|_0^1\right) = k\frac{(1-e^{-1})}{6} = 1$$

Thus:

$$f_{XYZ}(x, y, z) = \frac{6xy^2e^{-z}}{1-e^{-1}} \quad 0 < x, y, z < 1$$

To prove independence, we first need to calculate the marginal and joint cdfs.

$$F_X(x) = \int_0^1\int_0^1\int_0^x \frac{6\xi y^2e^{-z}}{1-e^{-1}}d\xi dydz = \frac{6}{1-e^{-1}}\left(\frac{\xi^2}{2}\Big|_0^x\right)\left(\frac{y^3}{3}\Big|_0^1\right)\left(-e^{-z}\Big|_0^1\right) = x^2$$

$$F_Y(y) = \int_0^1\int_0^y\int_0^1 \frac{6x\xi^2e^{-z}}{1-e^{-1}}dxd\xi dz = \frac{6}{1-e^{-1}}\left(\frac{x^2}{2}\Big|_0^1\right)\left(\frac{\xi^3}{3}\Big|_0^y\right)\left(-e^{-z}\Big|_0^1\right) = y^3$$

$$F_Z(z) = \int_0^z\int_0^1\int_0^1 \frac{6xy^2e^{-\xi}}{1-e^{-1}}dxdyd\xi = \frac{6}{1-e^{-1}}\left(\frac{x^2}{2}\Big|_0^1\right)\left(\frac{y^3}{3}\Big|_0^1\right)\left(-e^{-\xi}\Big|_0^z\right) = \frac{1-e^{-z}}{1-e^{-1}}$$

$$F_{XY}(x, y) = \int_0^1\int_0^y\int_0^x \frac{6\xi\varepsilon^2e^{-z}}{1-e^{-1}}d\xi d\varepsilon dz = \frac{6}{1-e^{-1}}\left(\frac{\xi^2}{2}\Big|_0^x\right)\left(\frac{\varepsilon^3}{3}\Big|_0^y\right)\left(-e^{-z}\Big|_0^1\right)$$

$$= x^2y^3$$

179

$$F_{XZ}(x, z) = \int_0^z \int_0^1 \int_0^x \frac{6\xi y^2 e^{-\varepsilon}}{1 - e^{-1}} d\xi dy d\varepsilon = \frac{6}{1 - e^{-1}} \left(\frac{\xi^2}{2} \Big|_0^x \right) \left(\frac{y^3}{3} \Big|_0^1 \right) \left(-e^{-\varepsilon} \Big|_0^z \right)$$

$$= x^2 \frac{1 - e^{-z}}{1 - e^{-1}}$$

$$F_{YZ}(y, z) = \int_0^z \int_0^y \int_0^1 \frac{6x\xi^2 e^{-\varepsilon}}{1 - e^{-1}} dx d\xi d\varepsilon = \frac{6}{1 - e^{-1}} \left(\frac{x^2}{2} \Big|_0^1 \right) \left(\frac{\xi^3}{3} \Big|_0^y \right) \left(-e^{-\varepsilon} \Big|_0^z \right)$$

$$= y^3 \frac{1 - e^{-z}}{1 - e^{-1}}$$

$$F_{XYZ}(x, y, z) = \int_0^z \int_0^y \int_0^x \frac{6\lambda\xi^2 e^{-\varepsilon}}{1 - e^{-1}} d\lambda d\xi d\varepsilon = \frac{6}{1 - e^{-1}} \left(\frac{\lambda^2}{2} \Big|_0^x \right) \left(\frac{\xi^3}{3} \Big|_0^y \right) \left(-e^{-\varepsilon} \Big|_0^z \right)$$

$$= x^2 y^3 \frac{1 - e^{-z}}{1 - e^{-1}}$$

Now to test for independence. By inspection, we can see that

$$F_X(x)F_Y(y) = F_{XY}(x, y), \quad F_X(x)F_Z(z) = F_{XZ}(x, z),$$

$$F_Y(y)F_Z(z) = F_{YZ}(y, z) \quad \text{and} \quad F_X(x)F_Y(y)F_Z(z) = F_{XYZ}(x, y, z). \text{ There-}$$

fore, X, Y, and Z are statistically independent.

Q.2 We start off by first determining the relationship for Z. In this case, for a differential amplifier we have Z=X-Y. Since X and Y can be assumed to be independent, [216] can be used:

$$f_Z(z) = \int_{-\infty}^{\infty} f_Y(y)f_X(z - y)dy$$

However, [216] was developed for X=Z-Y. In our current situation, X=Z+Y, therefore:

$$f_Z(z) = \int_{-\infty}^{\infty} f_Y(y)f_X(z + y)dy$$

180

The range of Z in our case is:

Zmin = Xmin - Ymax = -1 - 4 = -5

Zmax = Xmax - Ymin = 1 - (-3) = 4

Figure 95 below shows the condition for z>4. In this case $f_Y(y)f_X(z+y)=0$. Therefore:

$$f_Z(z) = 0 \; z>4$$

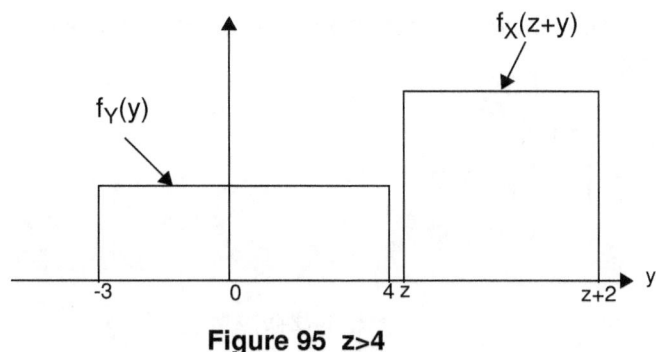

Figure 95 z>4

Figure 96 below shows the condition for $2 \leq z \leq 4$. In this case $f_Y(y)f_X(z+y) = 1/14$. Therefore:

$$f_Z(z) = \int_z^4 \frac{1}{14}dy = \frac{4-z}{14} \; 2 \leq z \leq 4$$

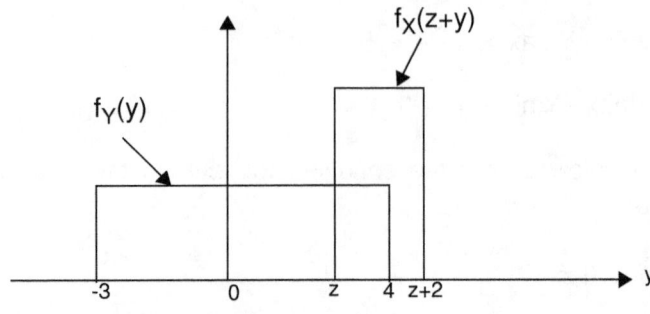

Figure 96 $2 \leq z \leq 4$

Figure 97 below shows the condition for $-3 \leq z \leq 2$. In this case $f_Y(y)f_X(z+y) = 1/14$. Therefore:

$$f_Z(z) = \int_{z}^{z+2} \frac{1}{14}dy = \frac{1}{7} \quad -3 \leq z \leq 2$$

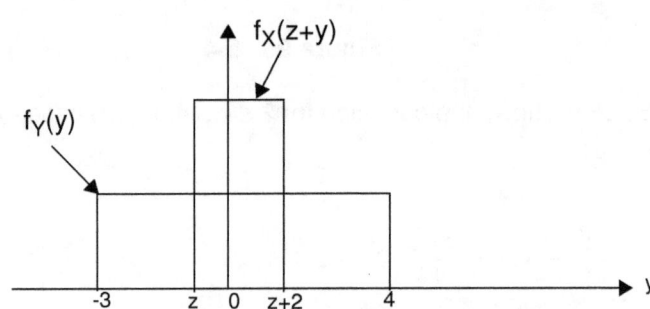

Figure 97 $-3 \leq z \leq 2$

Figure 98 below shows the condition for $-5 \leq z \leq -3$. In this case $f_Y(y)f_X(z+y) = 1/14$. Therefore:

$$f_Z(z) = \int_{-3}^{z+2} \frac{1}{14}dy = \frac{z+5}{14} \quad -5 \leq z < -3$$

182

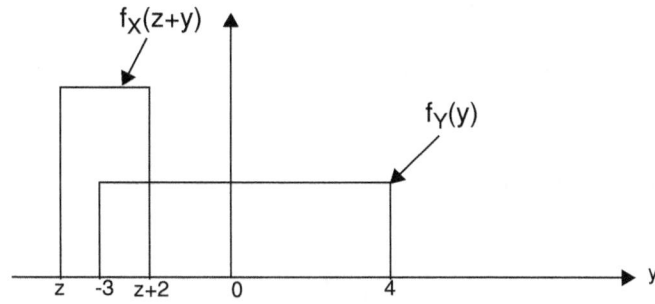

Figure 98 $-5 \le z < -3$

For the case where z<-5 $f_Z(z)=0$ because the two marginal pdfs do not overlap. Therefore:

$$f_Z(z) = \begin{cases} 0 & z < -5 \\ \dfrac{z+5}{14} & -5 \le z < -3 \\ \dfrac{1}{7} & -3 \le z \le 2 \\ \dfrac{4-z}{14} & 2 \le z \le 4 \\ 0 & z > 4 \end{cases}$$

There are two ways to obtain the mean of Z:

1. The easy way: $\overline{Z} = \overline{X}\text{-}\overline{Y} = 0\text{-}0.5 = \text{-}0.5$

2. The less easy way:

$$E[Z] = \int_{-5}^{-3} z\frac{(z+5)}{14}dz + \int_{-3}^{2} z\left(\frac{1}{7}\right)dz + \int_{2}^{4} z\frac{(4-z)}{14}dz = -\frac{1}{2}$$

Both methods agree, which is also a verification of the pdf.

Q.3 Using the characteristic function definition:

$$\Phi_X(\omega) = \int_{-\infty}^{\infty} \frac{e^{-\frac{\left(\frac{x-\bar{X}}{\sigma}\right)^2}{2}}}{\sqrt{2\pi}\sigma} e^{j\omega x} dx$$

Then making the substitution:

$$\left(u = \frac{x-\bar{X}}{\sigma}\right) \rightarrow \sigma du = dx \text{ and } x = u\sigma + \bar{X}$$

We get:

$$\Phi_X(\omega) = \int_{-\infty}^{\infty} \frac{e^{-\frac{u^2}{2}}}{\sqrt{2\pi}} e^{j\omega u\sigma + j\omega\bar{X}} du = \frac{1}{\sqrt{2\pi}} \int_{-\infty}^{\infty} e^{-\frac{u^2}{2} + j\omega u\sigma + j\omega\bar{X}} du$$

Using the following integral to arrive at the final result:

$$\int_{-\infty}^{\infty} e^{-au^2 + bu + c} du = \sqrt{\frac{\pi}{a}} e^{\frac{b^2 + 4ac}{4a}}$$

Thus:

$$\Phi_X(\omega) = \frac{1}{\sqrt{2\pi}} \left(\sqrt{2\pi} e^{\frac{(-\omega^2\sigma^2) + 4\left(\frac{1}{2}\right)j\omega\bar{X}}{2}} \right) = e^{-\frac{\omega^2\sigma^2}{2} + j\omega\bar{X}}$$

The mean is found by:

$$E[X] = \frac{1}{j}\frac{d}{d\omega}\Phi_X(\omega)\Big|_{\omega = 0} = \frac{-\omega\sigma^2 + j\bar{X}}{j} e^{-\frac{\omega^2\sigma^2}{2} + j\omega\bar{X}}\Big|_{\omega = 0} = \bar{X}$$

The mean square value is found by:

184

$$E[X^2] = \frac{1}{j^2}\frac{d^2}{d\omega^2}\Phi_X(\omega)\bigg|_{\omega=0} = \frac{1}{j^2}\left(e^{-\frac{\omega^2\sigma^2}{2}+j\omega\bar{X}}\right)((\omega\sigma^2-j\bar{X})^2-\sigma^2)\bigg|_{\omega=0}$$

$$= \frac{1}{j^2}[(j\bar{X})^2-\sigma^2] = \bar{X}^2+\sigma^2$$

The variance is then:

$$\sigma_X^2 = \overline{X^2}-\bar{X}^2 = \bar{X}^2+\sigma^2-\bar{X}^2 = \sigma^2$$

Q.4 First we find the pdf of Y:

$$f_Y(y) = \frac{f_X\left(\frac{y-2}{3}\right)}{\left|\frac{d}{dx}(3x+2)\right|} = \frac{1}{3} \quad -1\le y\le 2$$

Remember, ymin = 3(-1)+2=-1 and ymax = 3(0)+2=2.

Now we calculate the means and standard deviations:

$$E[Y] = E[g(x)] = \int_{-1}^{0} g(x)f_X(x)dx = \left(3\frac{x^2}{2}+2x\right)\bigg|_{-1}^{0} = \frac{1}{2}$$

$$E[X] = \frac{-1+0}{2} = -\frac{1}{2}$$

$$\sigma_X = \sqrt{\sigma_X^2} = \sqrt{\frac{(0-(-1))^2}{12}} = \sqrt{\frac{1}{12}} = \frac{1}{2\sqrt{3}}$$

$$\sigma_Y = \sqrt{\sigma_Y^2} = \sqrt{\frac{(2-(-1))^2}{12}} = \sqrt{\frac{9}{12}} = \frac{\sqrt{3}}{2}$$

$$E[XY] = E[X(3X+2)] = \int_{-1}^{0} 3x^2+2x\,dx = (x^3+x^2)\bigg|_{-1}^{0} = 0$$

$$\rho_{XY} = \frac{E[XY] - E[X]E[Y]}{\sigma_X \sigma_Y} = \frac{0 - \left(-\frac{1}{2}\right)\left(\frac{1}{2}\right)}{\left(\frac{1}{2\sqrt{3}}\right)\left(\frac{\sqrt{3}}{2}\right)} = 1$$

This result should not surprise us since Y and X are linearly related.

Q.5 From control system theory, we recognize that Y=AX/(1+A). Since Y is simply a function of X the mean is:

$$E[Y] = \frac{A}{1+A} \int_{-\infty}^{\infty} x f_X(x) dx$$

For a Gaussian random variable the integral evaluates to the mean of the Gaussian random variable, which is in this case equal to 0. Therefore, E[Y]=0. As an additional note, notice that the mean square value is:

$$E[Y^2] = \left(\frac{A}{1+A}\right)^2 \int_{-\infty}^{\infty} x^2 f_X(x) dx$$

Since the mean is zero the variance is equal to the mean square value. So even though the mean of the output signal is the same as that of the input signal[1], the variance of Y is multiplied by the square of the closed loop gain.

Q.6 The output of the transformer, let's call it S, will have a Gaussian pdf since the transformer only multiplies X by a constant factor. The mean and variance of S can be calculated as:

$$E[S] = \int_{-\infty}^{\infty} 2x f_X(x) dx = 2\bar{X} = (2)(0) = 0$$

$$\sigma_S^2 = \int_{-\infty}^{\infty} (2x - 2\bar{X})^2 f_X(x) dx = \int_{-\infty}^{\infty} 4(x - \bar{X})^2 f_X(x) dx = 4\sigma_X^2 = 4$$

---

1. Note that if the mean of X was non-zero its mean would be multiplied by the closed loop gain. This would also affect the variance.

186

Compare with [122] to see how the variance and mean are affected by the transformation.

Note that for S less than 0.7volts the diode will be reversed biased and the output will be equal to 0volts. Once S exceeds 0.7volts, the diode becomes forward biased and Y will equal S-0.7. There will be an impulse at Y=0 that will be equal to G((0-(-0.7))/2) = G(0.35). Using Table 2, G(0.35) = 0.6368. For Y≥0, we can use [122] with a=1 and b=-0.7. Remember that σ_S^2 =4.

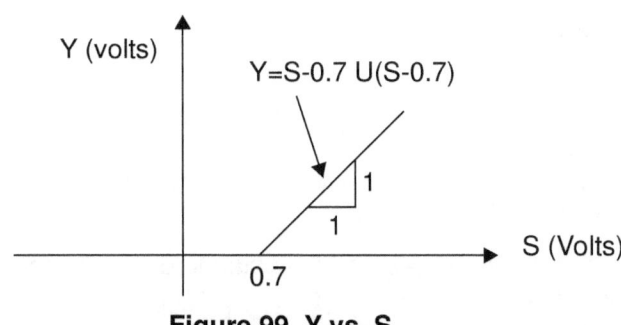

Figure 99 Y vs. S

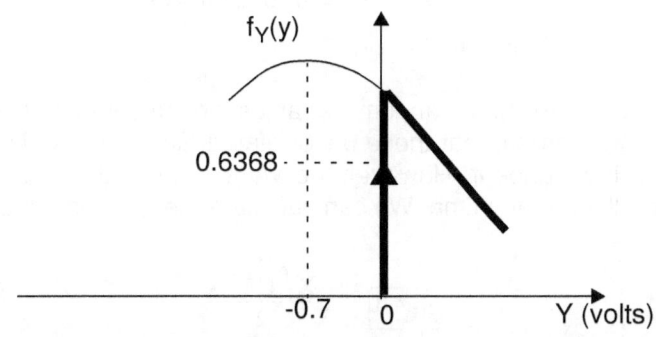

Figure 100 pdf of Y

So in this case we have to rely on [64] to find the mean. Therefore:

$$\bar{Y} = \int_{-\infty}^{\infty} y f_Y(y) dy = \int_{-\infty}^{\infty} 0.6368 y \delta(y) dy + \int_{0}^{\infty} y \frac{e^{-\frac{(y-(-0.7))^2}{2(2^2)}}}{\sqrt{2\pi}2} dy$$

$$= 0.49626 \text{ volts}$$

Q.7 The covariance matrix is defined as:

$$C = \begin{bmatrix} \sigma^2 x_1 & \rho_{x_1 x_2} \sigma_{x_1} \sigma_{x_2} & \rho_{x_1 x_3} \sigma_{x_1} \sigma_{x_3} \\ \rho_{x_2 x_1} \sigma_{x_2} \sigma_{x_1} & \sigma^2 x_2 & \rho_{x_2 x_3} \sigma_{x_2} \sigma_{x_3} \\ \rho_{x_3 x_1} \sigma_{x_3} \sigma_{x_1} & \rho_{x_3 x_2} \sigma_{x_3} \sigma_{x_2} & \sigma^2 x_3 \end{bmatrix}$$

Comparing with:

$$C = \begin{bmatrix} 0.25 & 0 & 0 \\ 0 & 5.00 & 0 \\ 0 & 0 & 0.05 \end{bmatrix}$$

and noting that the variances and thus standard deviations are non-zero, we can conclude that:

$$\rho_{X_n X_m} = \frac{E[X_n X_m] - E[X_n]E[X_m]}{\sigma_{X_n} \sigma_{X_m}} = 0 \text{ for all } n \neq m$$

This implies that the three random variables are uncorrelated. And as we have previously learned, that these uncorrelated Gaussian random variables are pair-wise independent. However, we want to also check their independence taken all three at a time. We can calculate the joint distribution as:

$$f_{X_1 X_2 X_3}(x_1, x_2, x_3) = \frac{\sqrt{|C^{-1}|}}{(2\pi)^{3/2}} e^{-\left(\frac{\mathbf{x}^T C^{-1} \mathbf{x}}{2}\right)}$$

$$C^{-1} = \begin{bmatrix} \dfrac{1}{0.25} & 0 & 0 \\ 0 & \dfrac{1}{5.00} & 0 \\ 0 & 0 & \dfrac{1}{0.05} \end{bmatrix}$$

$$\sqrt{|\mathbf{C}^{-1}|} = \left\| \begin{bmatrix} \dfrac{1}{0.25} & 0 & 0 \\ 0 & \dfrac{1}{5.00} & 0 \\ 0 & 0 & \dfrac{1}{0.05} \end{bmatrix} \right\| = \sqrt{\left(\dfrac{1}{0.25}\right)\left(\dfrac{1}{5}\right)\left(\dfrac{1}{0.05}\right)} = 4$$

$$\frac{\mathbf{X}^T \mathbf{C}^{-1} \mathbf{X}}{2} = \frac{\begin{bmatrix} x_1 & x_2 & x_3 \end{bmatrix} \begin{bmatrix} 4 & 0 & 0 \\ 0 & 0.2 & 0 \\ 0 & 0 & 20 \end{bmatrix} \begin{bmatrix} x_1 \\ x_2 \\ x_3 \end{bmatrix}}{2} = 2x_1^2 + 0.1x_2^2 + 10x_3^2$$

$$\therefore f_{X_1 X_2 X_3}(x_1, x_2, x_3) = \frac{4}{(2\pi)^{3/2}} e^{-(2x_1^2 + 0.1x_2^2 + 10x_3^2)}$$

Now note that the marginal probabilities are[1]:

$$f_{X_1}(x_1) = \frac{1}{(\sqrt{2\pi})(\sqrt{0.25})} e^{-(2x_1^2)} \,, \quad f_{X_2}(x_2) = \frac{1}{(\sqrt{2\pi})(\sqrt{5})} e^{-(0.1x_2^2)} \,,$$

$$f_{X_3}(x_3) = \frac{1}{(\sqrt{2\pi})(\sqrt{0.05})} e^{-(10x_3^2)}$$

By inspection, we can see that:

$$f_{X_1 X_2 X_3}(x_1, x_2, x_3) = \left(f_{X_1}(x_1)\right)\left(f_{X_2}(x_2)\right)\left(f_{X_3}(x_3)\right)$$

---

1. Just plug the mean and standard deviations into the formula:

$$f_X(x) = \frac{1}{\sqrt{2\pi}\sigma} e^{-\dfrac{\left(\dfrac{x - \overline{X}}{\sigma}\right)^2}{2}}$$

Which completes our proof that all three random variables are statistically independent.

Q.8 Since the currents are independent and identically distributed the central limit theorem and Table 2 can be applied here.

$$P(I \leq 1.6A) \ = \ G\left(\frac{1.6 - 12(0.12)}{\sqrt{12(0.2)}}\right) \ = \ G(0.10328) \cong 0.5517$$

$$P(I > 1.6A) \ = \ 1 - P(I \leq 1.6A) \cong 0.4483$$

Chapter 8　Bibliography

Below is a list of books which the reader can refer to if a more broad and detailed study into the topics presented is desired.

[1] Peebles, P. Z., Jr., (2000): *Probability, Random Variables, and Random Signal Principles*, 4th Edition, Tata McGraw-Hill, New Delhi, Print.

[2] Hsu, H. (1997): *Schaum's Outline of Theory and Problems of Probability, Random Variables, and Random Processes*, McGraw-Hill, New York, Print.

[3] Papoulis, A., and S. U. Pillai (2002): *Probability, Random Variables, and Stochastic Processes*, 4th Edition, Tata McGraw-Hill, New Delhi, Print.

[4] Cooper, G. R., and C. D. McGillem (1999): *Probabilistic Methods of Signal and System Analysis*, 3rd Edition, Oxford University Press, New York, Print.

[5] Bertsekas, D. P., and J. N. Tsitsiklis (2002): *Introduction to Probability*, 3rd printing, Athena Scientific, Belmont, Print.

[6] Brown, P., and P. Y. C. Hwang (1992): *Introduction to Random Signals and Applied Kalman Filtering*, 2nd Edition, Wiley, New York, Print.

[7] Davenport, W. B., Jr., and W. L. Root (1958): *An Introduction to the Theory of Random Signals and Noise*, McGraw-Hill, New York, Print.

[8] Lathi, B. P., (1998): *Modern Digital and Analog Communication Systems*, 3rd Edition, Oxford University Press, New York, Print.

[9] Garcia, A. L., (1989): *Probability and Random Processes for Electrical Engineering*, 2nd Edition, Addison-Wesley, USA, Print.

[10] Korn, G. A., and T. M. Korn (2000): *Mathematical Handbook for Scientists and Engineers: Definitions, Theorems, and Formulas for Reference and Review*, Dover, Mineola, Print.

[11] Kreyszig, E., (1999): *Advanced Engineering Mathematics*, 8th Edition, John Wiley & Sons, New York, Print.

[12] Pettofrezzo, Anthony J., (1978): Matrices and Transformations, Dover, New York, Print.

Index

www.ingramcontent.com/pod-product-compliance
Lightning Source LLC
Chambersburg PA
CBHW071424170526
45165CB00001B/386